North Korea's Weapons Programmes
a net assessment

D1194619

8 4·1
.K6
N67
2004

Preface

The International Institute for Strategic Studies (IISS) sees as one of its core missions the presentation to a wide public of the best available information on military holdings and strategy worldwide. For nearly half a century, the IISS has sought to provide facts on which intelligent policy analyses could be based. Each year we publish *The Military Balance*, an assessment of the military inventories held by some 170 countries. Especially since 2001, we have strengthened the information and analysis on opposition armies and terrorist groups. Given the greater saliency of transnational threats, there is a demand for more information on the military capabilities of non-state actors and also on their possible links with state actors. In addition, since 2002, we have strengthened our commitment and our abilities to provide information on nuclear, biological and chemical (NBC) weapons proliferation, as well as ballistic missiles.

Over this period, the Institute has developed a reputation for consistently providing an objective record of world-wide military holdings. The IISS has always applied careful judgements in analysing material in the public domain, and material offered to it by governments, international organisations and NGOs, individual experts with specialised knowledge, defectors from repressive states, journalists and others who have volunteered information or from whom we have sought comments on drafts of our work.

We are well aware of the different political agenda that sources and informants of all kinds may have in providing us with information or in commenting on information that we propose to publish or that we have published. The Institute tries carefully to discount for possible bias and to be exacting in the nature of the evidence that it chooses to rely on in presenting this to a wider public. Equally, in assessing the military holdings of non-state groups or of closed societies, reliable raw data is not easy to find. This is true for sophisticated intelligence agencies, international bodies, journalists, academics, and organisations like the IISS. In these cases it is important to explain the bases on which a more speculative judgement might have to be based.

During the Cold War, the IISS did its best to publish information on the East-West balance. This was sometimes contested by governments, given the political sensitivities attached to the slightest presumed change in the 'relationship of forces'. Often, changes that we recorded in *The Military Balance* were not a result of actual changes in force holdings, but were based on better information, or a reassessment of the old. This continues to this day, as we make clear in that publication, which we hope is improved and more accurate each year. Today, the great controversies surround the military capacities of non-state groups and the NBC holdings of proliferating states. This latter category is especially difficult, and inspires the most heated debates.

In September 2002, the IISS published *Iraq's Weapons of Mass Destruction: A Net Assessment.*

In that document, we attempted to bring together the best information available at the time, before inspectors were able to return to Iraq, on the history of Iraq's weapons programmes and the capacity that the regime had in a number of different areas to regenerate capabilities, accelerate work, renew production, or to add to stocks already held. Each chapter explained the nature of Iraq's historic ambitions and accomplishments in various programme areas. We explained, as clearly as could be done, the status of these programmes as understood by the international community and the inspectors that it had charged with investigating Iraqi activity in this area until they left Iraq in 1998. We then explained what Iraq might have been able to accomplish in each of the relevant programme areas. We then made a judgement as to the most likely current state of Iraq's programmes and inventory. We published a net assessment in tabular format. In each of the chapters from which these overall conclusions were drawn, we emphasised the range of possible outcomes. Whenever necessary we used the conditional tense in qualifying our judgements. We presented evidence, explained what the indications were in the absence of direct evidence, and where there was no clear evidence or strong indications we explained whether we thought a particular activity, programme or holding could or could not be ruled out.

We concluded that document by pointing out that 'This *Strategic Dossier* does not attempt to make a case, either way, as to whether Saddam Hussein's WMD arsenal is a *casus belli per se.*' We went on to say that: 'This *Strategic Dossier* invites policymakers and the public to make an early assessment of the relative risks of these different options and to choose a course that has the best chance of promoting regional and international security.' This wording was carefully chosen to underline the point that the IISS did not purport, in publishing the *Dossier*, to advance any specific policy aim.

This followed a longstanding Institute practice. The IISS does its best to separate any policy recommendations it may develop, or which we choose to publish by independent experts, from its presentation of the best available facts and assessments of military holdings.

Following the resumption of inspections in Iraq and then the war, it is clear that some of our assessments were proved correct, such as the development by Iraq of *al-Samoud* missiles that breached UN-set limits and our estimate that Iraq was at least several years away from being able to produce weapons-usable nuclear material or long-range ballistic missiles by itself. Other assessments, like the judgement that Iraq would probably deploy chemical artillery in the event of an attack, proved wrong. Some of our other judgements could still be proved right or wrong by the work of the Iraq Survey Group. Other assessments might never be tested against precise fact unless documents emerge or other findings come to light that can clearly explain what the position might have been in September 2002 and how that position might have been affected by measures taken by the regime of Saddam Hussein as the threat of war against him strengthened.

Overall, the IISS believes that *Iraq's Weapons of Mass Destruction: A Net Assessment* stands up well when compared with other documents published at about the same time both by governments and NGOs. Nevertheless, we intend to publish later in 2004 an independent assessment of that study. We will also devote efforts to analysing the intelligence challenges in the proliferation field.

In the meantime, there remains a need for independent organisations such as the IISS to make the best possible dispassionate efforts to analyse the proliferation challenges that confront the international community. This has been rightly identified as a key security threat. Our role must be to gather all the information available on these challenges and put them in the public domain. By doing so, we can explain the bases on which governments and other actors should be developing policy. By drawing together in one place not just information, but an explanation of why that information may be incomplete or suspect, we can make a key contribution to the debate.

That sense of mission has brought us to apply our investigative and analytical efforts to the issue of North Korea's weapons programmes.

For many reasons, the task of assessing North Korea's weapons programmes is even harder than that of judging those of Saddam Hussein's Iraq. Based on public information, conclusions about North Korea's nuclear, chemical and biological weapons capabilities, ballistic missiles programme and conventional forces are woven from a cloth of different strands. One strand of information is official reports from governments seeking to penetrate North Korea's veil of secrecy, such as public reports from the US, South Korea and Russia. However, this information is qualified. North Korea is a notoriously 'hard target' for intelligence collection. Reliable human sources are sparse, communication intercepts are fragmentary, and satellite and other remote sensing means provide limited information, subject to multiple interpretations. Governments must also be cautious that information released in public does not jeopardise already fragile sources and methods, assisting more effective North Korean concealment and deception efforts.

Given the difficulties of collecting information, government 'assessments' of North Korea are analytical judgements, based on evaluations and estimations of capabilities and motivations rather than hard conclusions based on conclusive evidence. A North Korean intelligence analyst could safely 'assert' that the United States has nuclear weapons and South Korea does not. The comparable American analyst may only be able to 'assess' that North Korea has nuclear weapons. Conclusive proof is lacking and Pyongyang likes to keep the world guessing. In assessing ambiguous and uncertain foreign threats, intelligence agencies naturally lean towards 'worst case' assessments to err on the side of caution. This is a universal and understandable tendency. Prudently, governments prefer to plan on the worst case rather than be surprised by it.

Another strand of information on North Korea comes from direct observations. For nearly a decade, North Korea's main nuclear facilities at the Yongbyon nuclear centre have been inspected and monitored by the International Atomic Energy Agency (IAEA), which provides a technological baseline for evaluating their capabilities. Observation of North Korean missile tests provides some technical characteristics and parameters of missiles under development, and periodic interdictions of missile-related exports provide technical details of missiles that are being produced. This information is also limited. IAEA inspections have not extended to clandestine facilities, such as those presumed to exist in association with North Korea's enrichment programme, or to facilities and activities linked to nuclear weapons development, which are beyond the mandate of IAEA inspectors. Observations of missile tests and interception of missile exports cannot illuminate less observable research and development activities or answer questions of overall North Korean missile production, deployment and armament.

In addition to official government reports and direct observation, there are a number of weaker strands of information. Purported 'leaks' of government intelligence may provide insight into sensitive information, but it must be treated with caution. Leaked information is not necessarily accurate information. Officials who disclose classified intelligence to the media may have a political agenda, and it is often difficult to verify the veracity of such information. Over the last decade, a number of North Korean defectors and refugees have come forward with intriguing information on North Korean military programmes. Some of this information has proved to be credible, some of it is implausible, and some cannot be confirmed. Distinguishing the wheat from the chaff is often impossible, even for government agencies that seek to glean insights from such sources. As always,

Preface

information from defectors and refugees has to be weighted against the possibility that they are unintentionally passing on false secondhand accounts or even fabricating tales for political and financial gain. To help illustrate the kind of information provided by such defectors, we have included tables in some of the chapters summarising the most important defectors that are publicly known and the information they have provided. There are also a number of secondary sources – press stories, magazine articles, reports from research organisations, books, and so forth. We have reviewed these for additional information and analysis, and cited the most valuable of these for readers who wish to explore subjects further.

Finally, there is information from the North Korean government itself. Obviously, this information has to be measured against Pyongyang's interest in manipulating the outside world's perceptions of its capabilities. At times, this interest may dictate denial of capabilities that actually exist. At other times, national interest may dictate invention or exaggeration of capabilities that do not exist. Moreover, North Korean efforts to sway and shape international perceptions are more sophisticated than public announcements from government agencies and private confidences from officials. North Korea is aware of the ways in which the outside world seeks to penetrate its secrets, and it takes active measures to mislead and conceal, whether to hide real capabilities or to create an impression of capabilities that do not exist.

With all these pitfalls in mind, we have tried to present a balanced and cautious set of assessments in individual chapters on North Korea's nuclear programme, its chemical and biological programmes, its ballistic missile programme, and the conventional military balance on the Korean Peninsula. To help establish political context, we have also included an opening chapter that recounts nearly 25 years of diplomatic efforts to deal with the North Korean nuclear and ballistic missile issues. The conclusion seeks to summarise the results of our efforts, identifying the underlying assumptions about technical capabilities and political motivations that form the basis for our judgements. To produce this *Dossier*, we used the same process followed for Iraq. Recognised technical experts were invited to draft chapters on each of the areas, and the chapters were then subject to review and comment by a wide range of experts in the field. The editor, aided by the research assistants, then produced a new draft, incorporating comments and additional information, and the original authors were given a final opportunity to respond to the penultimate drafts. The editor thanks the IISS editorial and design staff for their contribution to this publication.

Finally, we thank the various individuals who have contributed their knowledge and experience to the compilation of this *Dossier*. The responsibility for the information and judgements that we present is, unambiguously, ours alone.

Disarmament Diplomacy with North Korea

Overview

Diplomatic efforts to deal with North Korea's programmes to acquire nuclear weapons and develop its ballistic missile capabilities have, over nearly 25 years, witnessed both success and failure. During this period, four different approaches have been tried. Firstly, beginning in the 1980s, the US led efforts to employ pressures and inducements to convince North Korea to adhere to the 1968 Treaty on the Non-proliferation of Nuclear Weapons (the nuclear Non-proliferation Treaty, or NPT) and accept International Atomic Energy Agency (IAEA) inspection of its nuclear facilities and materials. In December 1985, North Korea acceded to the NPT and, after significant prevarication, accepted international inspections in April 1992. However, implementation of the inspection agreement collapsed when North Korea refused to cooperate with the IAEA to verify plutonium production prior to 1992, a situation compounded when Pyongyang threatened to withdraw from the NPT in March 1993.

Secondly, in December 1991, North and South Korea reached agreement on a bilateral 'denuclearisation' agreement that included restrictions on nuclear activities beyond those specified in the NPT. But this agreement was never implemented due to disagreements between Seoul and Pyongyang over the number and type of bilateral inspections necessary to verify it. The North–South agreement remained a dead letter until it was officially renounced by Pyongyang in May 2003.

Thirdly, following North Korea's threat to withdraw from the NPT in March 1993, the US and North Korea in October 1994 concluded a bilateral agreement known as the Agreed Framework. The Agreed Framework called for an ambitious undertaking to freeze and eventually dismantle North Korea's plutonium production facilities and account for its plutonium stocks in exchange for interim supplies of heavy fuel oil and an alternative nuclear energy project, as well as improved bilateral relations with Washington. For nearly a decade, the Agreed Framework froze North Korean production of additional plutonium, but it did not end North Korea's efforts to acquire nuclear weapons. Following public revelations in October 2002 that North Korea was pursuing a secret programme to produce weapons-grade uranium, the Agreed Framework collapsed amid diplomatic acrimony. North Korea revived its plutonium production facilities in December 2002 and withdrew from the NPT in January 2003, arguing that it had already given the requisite 90-day notice when it announced its original intent to withdraw in March 1993.

Finally, since the collapse of the Agreed Framework, the US has promoted a fourth effort to deal with the North Korean nuclear issue – through Six Party Talks between the US, Russia, China, Japan, and North and South Korea. These are intended to secure a multilateral agreement for North Korea to abandon its nuclear weapons programme in exchange for security assurances and political and economic benefits. The outcome of the Six Party Talks is uncertain.

Early pressure and engagement (1980–92)

In 1980, US intelligence detected the construction of a new research reactor at North Korea's Yongbyon Nuclear Research Centre, which US experts concluded could be designed to produce plutonium for a nuclear weapons programme. At the time, Washington responded by urging the Soviet Union to convince North Korea to join the NPT and allow IAEA inspections. Despite their Cold War rivalry, Washington and Moscow maintained regular talks on non-proliferation issues in order to coordinate policies and share intelligence on proliferation threats. Following Moscow's intervention, Pyongyang acceded to the NPT on 12 December 1985. In return, Moscow promised to sell North Korea four light water reactors (LWRs) for the generation of nuclear energy.

While North Korea's accession to the NPT was seen in Washington as a victory for its strategy of applying indirect pressure through Moscow, celebrations proved premature. Under the terms of the NPT, Pyongyang was – within 18 months – required to ratify and implement a 'full-scope safeguards' agreement with the IAEA. This required North Korea to declare all nuclear materials and facilities in the country to the IAEA and allow the agency's inspectors to verify that these were being used for peaceful purposes. However, Pyongyang found a variety of reasons to procrastinate. At first, North Korea blamed the IAEA, which had sent the wrong documentation to Pyongyang and did not discover its mistake for 18 months. Later, Pyongyang sent messages through Moscow to the effect that a reduction of tension in the US–North Korean relationship would 'facilitate' Pyongyang's completion of a safeguards agreement. But Washington refused to accept any linkage between the status of bilateral US–North Korean relations and North Korea's compliance with legal obligations under the NPT. Publicly, North Korea denied that it possessed any undeclared nuclear facilities and frequently charged that US nuclear weapons stationed in South Korea posed a grave threat to its country.

In the meantime, North Korea proceeded with its indigenous nuclear programme. The new reactor, now

---

'While North Korea's accession to the NPT was seen in Washington as a victory for its strategy of applying indirect pressure through Moscow, celebrations proved premature'

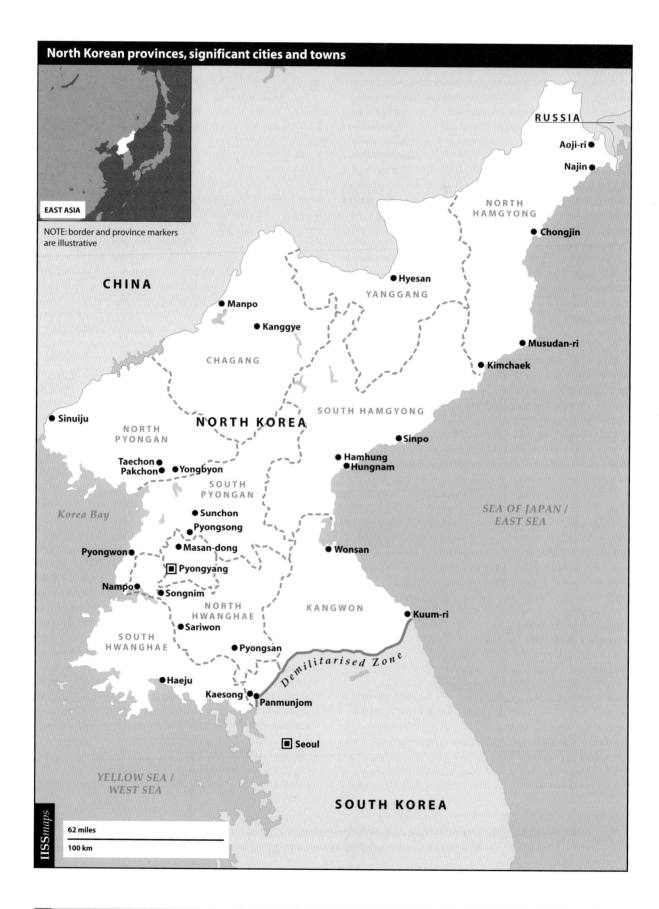

North Korean provinces, significant cities and towns

EAST ASIA

NOTE: border and province markers are illustrative

CHINA

RUSSIA

Aoji-ri ●

Najin ●

NORTH HAMGYONG

● Chongjin

● Hyesan

● Manpo

YANGGANG

● Kanggye

CHAGANG

● Musudan-ri

● Kimchaek

SOUTH HAMGYONG

● Sinuiju

NORTH PYONGAN

NORTH KOREA

● Sinpo

Taechon ●
Pakchon ● ● Yongbyon

● Hamhung
● Hungnam

SOUTH PYONGAN

Korea Bay

SEA OF JAPAN / EAST SEA

● Sunchon
● Pyongsong

Pyongwon ●
● Masan-dong

● Wonsan

Nampo ●
▣ Pyongyang

● Songnim

NORTH HWANGHAE

KANGWON

● Kuum-ri

SOUTH HWANGHAE
● Sariwon

● Pyongsan

Demilitarised Zone

● Haeju

Kaesong ●
● Panmunjom

▣ Seoul

YELLOW SEA / WEST SEA

SOUTH KOREA

62 miles

100 km

IIS*Smaps*

known as the 5MW(e) graphite-moderated reactor, began operations in 1986. By the end of the decade, construction had begun on an even larger reactor – now known as the 50MW(e) graphite-moderated reactor – with significantly greater capacity to produce plutonium. In addition, North Korea had completed a number of new facilities at Yongbyon, one of which, some US experts believed, might be a reprocessing plant for separating plutonium from irradiated fuel rods. Concerned by these advances, the incoming administration of President George H.W. Bush decided to pursue a more activist policy, offering to improve relations with Pyongyang if it fulfilled its NPT safeguards commitments and refrained from pursuing enrichment and reprocessing technologies. At the same time, Washington adopted a balanced policy of promoting improvements in North–South relations while maintaining a strong military deterrent on the Peninsula.

During the summer of 1990 and spring of 1991, Washington's attention was focused on ejecting Iraqi forces from Kuwait, and little effort was made to implement a new policy towards North Korea. Towards the end of 1991, however, the collapse of the Soviet Union weakened North Korea's strategic position by depriving it of a source of aid and political support, and created an opening for US diplomatic efforts to resolve the North Korean nuclear issue. On 27 September 1991, President Bush announced that all land- and sea-based US tactical nuclear weapons would be removed from overseas locations. This included all US nuclear weapons in South Korea. Although the announcement was primarily intended to achieve a parallel withdrawal of Soviet tactical nuclear weapons, it also gave Pyongyang an opportunity to declare that the nuclear threat from US forces in South Korea was removed. Following the Bush initiative, South Korean President Roh Tae Woo declared that his country was nuclear-free and proposed talks with Pyongyang to establish a nuclear-free zone on the entire Korean Peninsula. On 31 December 1991, Seoul and Pyongyang announced the North-South Denuclearization Declaration (NSDD), which banned the development and possession of nuclear weapons as well as enrichment and reprocessing facilities, and called for a North–South inspection regime to verify the agreement. In private, Washington had advised Seoul on the substance of the NSDD. Many in Washington saw this regional approach as a means of overcoming limitations in the NPT (which did not ban reprocessing and enrichment), as well as bolstering an IAEA

inspection system that had failed to detect Iraq's massive clandestine nuclear weapons programme before the 1991 Gulf War.

By early 1992, there was a parallel breakthrough on the IAEA safeguards impasse. In January 1992, during a visit to Seoul by President Bush, the United States and South Korea pledged to suspend their annual *Team Spirit* military exercise on the condition that North Korea concluded its inspection agreement with the IAEA. Later that month, at an historic meeting between US Under Secretary of State Arnold Kanter and North Korean Party Secretary Kim Young Sun, Washington additionally offered to improve bilateral relations if Pyongyang fulfilled its non-proliferation obligations. On 30 January 1992, North Korea signed its full-scope safeguards agreement with the IAEA, which entered into force on 10 April 1992. In March, IAEA Director General Hans Blix was taken on a tour of the previously undeclared nuclear facilities at Yongbyon, and in May formal IAEA inspections got underway to verify the accuracy of North Korea's initial declaration of nuclear facilities and materials.

Prior to the IAEA inspections, Washington's knowledge of Yongbyon was largely limited to analysis gleaned from satellite intelligence, and there were many uncertainties about the status and capabilities of the facilities being monitored. The initial IAEA inspections revealed that the North Korean programme was more advanced than the US had previously assessed. In particular, North Korea declared to the IAEA that it had shut down the 5MW(e) reactor for a few months in 1989–90 to remove a few hundred damaged fuel rods, some of which it had reprocessed to recover a small amount of plutonium. Over the summer of 1992, IAEA analysis of samples taken at the reprocessing facilities suggested that North Korea had produced more plutonium than it had declared, although the sample analysis could not determine the amount. Adding to suspicions, the US detected apparent North Korean efforts to conceal from the IAEA two underground sites, which could contain waste from undeclared reprocessing. Analysing the operational history of the 5MW(e) reactor – as well as a small Soviet-supplied research reactor at Yongbyon – US experts calculated that before 1992, North Korea could have discharged and reprocessed enough additional spent fuel to extract plutonium for one or possibly two nuclear weapons. This assessment represented a worst case scenario. To determine the actual amount of undeclared plutonium, the IAEA would need to measure this nuclear waste, presumably stored at the two suspect nuclear waste sites.

'The initial IAEA inspections revealed that the North Korean programme was more advanced than the US had previously assessed'

Towards the end of 1992, tensions increased between the IAEA and Pyongyang as the agency sought access to the suspect waste sites, and North Korea responded that the sites were 'military-related' and beyond the IAEA's jurisdiction. North–South relations also deteriorated as negotiations to implement the NSDD stalled over the type and frequency of inspections that North Korea would accept. In October 1992, the United States and South Korea announced that they would continue preparations for the 1993 *Team Spirit* exercise in the absence of progress on the bilateral inspection regime. Washington subsequently rejected North Korean requests for another high-level meeting, until progress was made in implementing North Korea's IAEA safeguards agreement and the NSDD.

Eruption of the North Korean nuclear crisis

As President Bill Clinton took office in January 1993 and South President Kim Young Sam took office in February, both new administrations inherited an impending nuclear crisis with North Korea. Having exhausted all voluntary measures to gain access to the suspect waste sites, the IAEA made a formal request in late January 1993 for a 'special inspection', invoking a provision in the safeguards agreement that allowed the IAEA to seek access to any site it believed was necessary to verify safeguards. When further consultations with North Korea failed to produce any progress, a special meeting of the IAEA Board of Governors was held in late February, at which the US displayed satellite imagery showing the suspect nuclear waste sites. Despite their reluctance to confront North Korea, Russia and China joined the US and other countries in adopting a unanimous Board of Governors Resolution on 25 February 1993, giving North Korea a deadline of one month to comply with its safeguards obligations. By coincidence, the *Team Spirit* exercise also began in February.

On 12 March 1993, in the face of mounting international pressure, North Korea suddenly announced that it intended to withdraw from the NPT, invoking a clause in the Treaty that allows a party to pull out in consideration of its 'supreme interests' after giving 90 days notice. Pyongyang said it was being forced to withdraw because of US military threats as well as efforts to manipulate the IAEA to gain access to military sites. But North Korea also suggested that it would reconsider withdrawal if 'the United States stops its nuclear threats against our country and the IAEA returns to the principles of independence and impartiality'. Surprised by North Korea's

announcement, Washington responded in two ways. Firstly, the US sought to increase diplomatic pressure and threaten sanctions if North Korea did not back down. The US secured a 25 March 1993 IAEA Resolution that condemned North Korea and reported its violation of the NPT to the United Nations Security Council (UNSC). Secondly, the US decided to engage in direct talks with North Korea, calculating that diplomatic pressure and the threat of sanctions would not be sufficient to disarm Pyongyang. Moreover, other key players, like China, Russia, South Korea and Japan, made it clear that they would not support tougher measures unless Washington first tried to resolve the issue through dialogue with Pyongyang.

Both approaches were combined in UNSC Resolution 825, adopted on 11 May 1993, which called on North Korea to retract its threat to renounce the NPT and to honour its safeguards obligations, and urged 'interested states' to facilitate a solution. To obtain Russian and Chinese support for the resolution, as well as the backing of South Korea and Japan, Washington said that it was prepared to begin talks with North Korea and asked Beijing to pass a private message to Pyongyang that the US was willing to meet after the resolution passed. From Washington's perspective, the resolution provided helpful political cover to begin bilateral efforts with Pyongyang at the request of the international community, without appearing to reward North Korea's threats.

On the basis of Resolution 825, the US and North Korea entered into 17 months of difficult negotiations, which eventually produced the October 1994 Agreed Framework. Ambassador Robert Gallucci led the US team, an interagency delegation representing the National Security Council, the Departments of State and Defense, and the Joint Chiefs of Staff, while Vice Foreign Minister Kang Sok Ju headed a North Korean team from the Ministry of Foreign Affairs. Major rounds of talks were held in June 1993, July 1993, July 1994, August 1994, and September–October 1994. In these meetings, the two sides laid out their formal positions in large plenary sessions, usually reading from prepared scripts. The real negotiating was done in scores of smaller and less cumbersome informal sessions. In these, the heads of delegations, or smaller groups of experts, could explore possible compromises and draft language for the joint communiqués issued at the end of each round, and also negotiate the language of the Agreed Framework itself. In between official rounds, officials from Washington and North Korean officials from Pyongyang's UN mission held scores of

'The US decided to engage in direct talks with North Korea, calculating that diplomatic pressure and the threat of sanctions would not be sufficient to disarm Pyongyang'

meetings in New York (known as the 'New York channel'). Throughout the talks, the US remained in close contact with South Korea and Japan, usually briefing delegations from Seoul and Tokyo at the end of every day of talks with the North Koreans. The US team also made frequent trips to Seoul and Tokyo to consult with senior officials, as well as visiting Beijing, Moscow, and other capitals to seek support for US negotiating positions.

The first round of bilateral talks, held in New York in mid-June 1993, succeeded in 'suspending' North Korea's announced withdrawal from the NPT, which meant that IAEA inspectors would continue to monitor North Korea's nuclear activities while the talks proceeded. As a condition for the talks, Washington demanded that North Korea accept limits on its nuclear programme, such as refraining from additional reprocessing, while the talks were underway. In return, the US agreed to a joint document with North Korea pledging principles of non-aggression and non-interference – drawn from the United Nations Charter – which would govern the bilateral relationship. In the discussions, North Korea indicated it would remain a party to the NPT and fulfill its safeguards obligations if the US removed its 'nuclear threat' and ensured that the IAEA applied safeguards 'impartially', which, Pyongyang explained, meant withdrawing its request for special inspections of the suspect waste sites.

However, in the second round – held in Geneva in July 1993 – North Korea unveiled a more ambitious proposal to abandon its plutonium production programme in return for the provision of a light water reactor (LWR) project by the United States. This proposal was attractive to Washington because implementing it would prevent North Korea from accumulating a stockpile of separated plutonium under IAEA safeguards, as it was legally allowed to do under the NPT. With such a stockpile of separated plutonium, North Korea could 'break out' from the NPT and fabricate nuclear weapons in short order. Safeguards on the LWR project and provisions to remove its spent fuel from North Korea could guard against the danger that North Korea would be able to secretly extract plutonium from the LWR spent fuel. Nonetheless, the proposal had several drawbacks. Even if North Korea eschewed further plutonium production, there remained the issue of undeclared plutonium produced prior to 1992, potentially enough for one or two bombs. Moreover, the cost and practicality of the proposed LWR project were daunting. Clearly, the US could only undertake such a

project with massive investment by other countries, such as South Korea and Japan. In the July meeting's joint communiqué, the United States agreed to 'consider' the introduction of new reactors into North Korea, and the two sides agreed to meet again in September. Before then, at US insistence, North Korea was required to begin talks with the IAEA to arrange for new inspections of Yongbyon and with South Korea on implementation of the NSDD.

Ultimately, the US effort to simultaneously move ahead on three separate tracks (US–North Korea, North Korea–South Korea, and IAEA–North Korea) proved unworkable, and the planned September 1993 round of bilateral talks between Washington and Pyongyang had to be postponed. Finally, after numerous meetings in New York, US and North Korean representatives reached agreement on a package of four steps scheduled to take place on Tuesday, 1 March 1993 – dubbed 'Super Tuesday'. Under the agreement, North Korea would allow the IAEA to conduct an inspection at Yongbyon to ensure the 'continuity of safeguards'; North and South Korea would agree to exchange special envoys; the US and South Korea would announce cancellation of the 1994 *Team Spirit* exercise; and the US and North Korea would announce dates for the third round of talks. The 'Super Tuesday' deal collapsed, however, when North Korea refused to allow the IAEA to complete inspections at Yongbyon, and plans for a meeting of North–South envoys before the next round of US–North Korea talks fell apart, with mutual recriminations between Seoul and Pyongyang. Washington responded by cancelling the scheduled US–North Korea talks and renewing *Team Spirit* planning with Seoul. In May 1994, North Korea publicly warned that it intended to unload the nearly 8,000 spent fuel rods from the 5MW(e) reactor as a 'safety measure'. Despite US objections, North Korea proceeded to carry out its threat in June, unloading the spent fuel and refusing to allow the IAEA to record the location of the individual fuel rods in the core, thus destroying one of the key technical means for measuring the operational history of the reactor and determining total plutonium production. More ominously, removal of the 8,000 rods was the first step towards possible reprocessing, which could have begun a few months later once the rods – placed in a storage pond near the 5MW(e) reactor – had cooled down. According to IAEA estimates, the 8,000 rods contained about 25–30 kilograms (kg) of plutonium, notionally enough for a few simple fission-type nuclear weapons.

'*North Korea unveiled a more ambitious proposal to abandon its plutonium production programme in return for the provision of a light water reactor (LWR) project by the United States*'

June 1994 marked the highpoint of tensions. When North Korea unloaded the 5MW(e) reactor in the face of US protests, Washington made preparations to seek UNSC sanctions, although the support of China and Russia was uncertain. In the event that agreement could not be reached in New York, Washington consulted with Seoul and Tokyo about imposing ad hoc sanctions, such as cutting off remittances from Korean immigrants in Japan. Since Pyongyang had declared that it would consider imposition of UNSC sanctions to be an 'act of war,' the US Department of Defense presented plans to President Clinton to quickly reinforce US forces on the Peninsula. A pre-emptive strike to destroy the Yongbyon reprocessing plant was also under consideration. Into this gathering storm, former President Jimmy Carter travelled to Pyongyang in June for direct talks with North Korean leader, Chairman Kim Il Sung. Kim promised Carter that IAEA inspectors would be allowed to remain in North Korea to monitor nuclear activities. Washington 'interpreted' Kim's commitment to include a pledge not to reload the 5MW(e) with fresh fuel nor to reprocess the 8,000 spent fuel rods, and Pyongyang consented to these conditions as a basis for resuming negotiations. Chairman Kim Il Sung also agreed to a proposal relayed by Carter from South Korean President Kim Young Sam to hold the first-ever North–South summit.

Birth of the Agreed Framework
Once this 'freeze' was established, the United States and North Korea returned to the negotiating table in Geneva in July, but the talks were interrupted on 8 July 1994 by the death of Chairman Kim Il Sung. Following a mourning period, Kim Il Sung's son, Kim Jong Il, emerged as the new leader of North Korea, and the negotiations carried on with lengthy drafting sessions in August and September–October, finally culminating in the conclusion of the Agreed Framework on 21 October 1994. During these final negotiations, two issues emerged as deal-breakers. Firstly, the North Koreans strongly resisted the implementation of special inspections, which would have required them to reveal their existing plutonium cache. The North Korean negotiator, Kang Sok Ju, was fond of describing what unenviable fate would await him if he returned home having accepted special inspections. Secondly, North–South relations sharply deteriorated after the death of Kim Il Sung, in part because Pyongyang thought that South Korean President Kim Young Sam showed disrespect to the dead leader. As a result,

Pyongyang strongly opposed any effort to improve North–South relations as part of the agreement. For his part, unable to salvage a North–South summit, Kim Young Sam tried to delay a US–North Korea deal and link any agreement to progress on the North–South front. In order to ensure that Seoul would be a key player in any nuclear deal, Kim Young Sam pledged that his country would build and largely fund the LWR project in North Korea, which would be based on a power reactor developed by the South Korean nuclear industry, originally derived from an American design. In contrast to Seoul, Tokyo was generally comfortable with the outlines of the deal, although there was some nervousness about costs (Japan had pledged to be the second largest funder of the LWR project after South Korea).

Both issues – special inspections and North–South relations – were not resolved until the eleventh hour. In a private meeting with Ambassador Gallucci, Kang Sok Ju offered for the first time to accept special inspections, but only once a major part of the light water reactors had been completed. On the basis of this suggestion, experts from both sides worked out details to tie the scheduling of full-scope safeguards implementation – including special inspections – to the completion of a 'significant portion' of the LWR project (This was defined in a Confidential Minute to the Agreed Framework as the point at which the major non-nuclear components of the first LWR unit were completed, but before the delivery of key nuclear components.) Initially, President Kim Young Sam was publicly opposed to this compromise, but eventually came on board when the US was able to secure North Korean agreement to include references to North–South dialogue and NSDD implementation in the Agreed Framework. North Korea understood that the LWR project would consist of 'South Korean type' reactors, but – for reasons of face – they asked that this not be made explicit in the Agreed Framework. Later, the type of reactor would emerge as a major stumbling block in efforts to implement the Agreed Framework.

Throughout the negotiations, China and Russia were active on the sidelines and behind the scenes. Publicly, China supported the establishment of a nuclear-free Korean Peninsula through peaceful dialogue. Privately, Chinese officials said they had little influence over Pyongyang and urged the US to resolve the dispute through bilateral negotiations with North Korea. After the Agreed Framework was concluded, however, some Chinese officials claimed that they had helped to pressure North Korea into an agreement by warning

'A pre-emptive strike to destroy the Yongbyon reprocessing plant was also under consideration'

Pyongyang not to count on Chinese protection if the negotiations collapsed and Washington sought UNSC sanctions. Russia's role was relatively marginal. Because of its own economic difficulties, which forced an end to its assistance to North Korea, Moscow was thought to have limited influence with Pyongyang. In addition, some important elements of the Russian government resented the Agreed Framework, because they believed the LWR project had replaced Russia's earlier agreement to sell power reactors to North Korea.

Under the terms of the Agreed Framework and the Confidential Minute, North Korea immediately froze its 'graphite moderated reactors and related facilities' (the operational 5MW(e) reactor and the 50MW(e) and 200MW(e) reactors under construction, along with the fuel fabrication facility, and the reprocessing plant), which were placed under IAEA monitoring. In return, the US gave assurances – backed by a Presidential letter – that the US would 'organize under its leadership an international consortium to finance and supply the LWR project' and provide interim heavy fuel oil (HFO) supplies to North Korea for heating and electricity production. The timing and size of these HFO shipments were calibrated to the notional electrical output from the graphite-moderated reactors that North Korea was freezing, reaching a level of 500,000 tons of HFO annually after one year.

The Agreed Framework was structured to require North Korean disarmament in stages, linked to progress in the supply of the LWR project, which consisted of two 1,000MW(e) units to be completed by a 'target date of 2003'. In the first stage, North Korea froze additional plutonium production, but retained a residual nuclear weapons capability (i.e. whatever plutonium it had produced before 1992) until a 'significant portion' of the LWR project was completed, estimated to be at least four to five years away. At that point, North Korea was required to come into full compliance with its safeguards agreement with the IAEA, 'including taking all steps deemed necessary by the IAEA'. Only after the IAEA had completed its verification would shipment of key nuclear components for the LWR project begin. Thus, the Agreed Framework did not require immediate North Korean compliance to account for plutonium produced prior to 1992 – the issue that had sparked the crisis in the first place. To critics, the Agreed Framework undermined the NPT regime because it allowed North

Korea to remain in violation of its safeguards obligations and to retain a small amount of undeclared plutonium, perhaps enough for one or two nuclear weapons. To its supporters, the Agreed Framework accepted a delay in North Korean compliance with safeguards in exchange for achieving a freeze on North Korea's plutonium production programme which, if left unhindered, could have produced much larger quantities of plutonium – even under IAEA safeguards.

At further stages of the LWR project, North Korea was required to take additional steps to dismantle its nuclear capabilities. When the first unit of the LWR project was completed, the removal from North Korea of the 8,000 spent fuel rods (containing 25–30kg of plutonium) from the 5MW(e) reactor would also be completed. In the meantime, the US agreed to work with North Korean technicians to stabilise the rods in the spent fuel pond at Yongbyon, to ensure safe storage. Finally, the dismantlement of North Korean graphite-moderated reactors and related facilities would be finished on the completion of the second LWR reactor unit. Thus, the Agreed Framework set up two simultaneous and linked set of actions: as the LWR project was constructed, North Korea's plutonium production facilities would be dismantled, and vice versa.

In addition to these specific nuclear disarmament provisions, the Agreed Framework included more general language calling for steps to improve economic and political relations between Washington and Pyongyang. Within three months, the US promised to 'reduce barriers to trade and investment' and the two sides agreed to open liaison offices in each other's capitals and eventually upgrade bilateral relations to the ambassadorial level 'as progress is made on issues of concern to both sides.' The US also agreed to 'provide formal assurances to the DPRK [Democratic People's Republic of Korea – North Korea's official name], against the threat or use of nuclear weapons by the US' – although US negotiators explained that these assurances would not be provided until North Korea was verified as a non-nuclear weapons state by complying with its IAEA safeguards agreement under the NPT.

From Washington's perspective, the Agreed Framework's political and economic provisions were designed to build up incentives for Pyongyang to sacrifice its residual nuclear weapons capability once a

---

'To critics, the Agreed Framework undermined the NPT regime because it allowed North Korea to remain in violation of its safeguards obligations and to retain a small amount of undeclared plutonium, perhaps enough for one or two nuclear weapons' ... 'To its supporters, the Agreed Framework accepted a delay in North Korean compliance with safeguards in exchange for achieving a freeze on North Korea's plutonium production programme which, if left unhindered, could have produced much larger quantities of plutonium – even under IAEA safeguards'

Disarmament Stages in the 1994 Agreed Framework

Stage One: The Nuclear Freeze

The US will undertake to make international arrangements for the provision of a LWR project (consisting of two 1,000MW(e) units) to North Korea by a target date of 2003 and to provide heavy fuel oil to North Korea for heating and electricity production, to offset the energy foregone due to the freeze of North Korea's graphite-moderated reactors and related facilities, pending completion of the first LWR unit. North Korea will freeze its graphite-moderated reactors and related facilities and provide full cooperation to IAEA to monitor the freeze. The freeze will consist of no refuelling or operation of the 5MW(e) reactor, freezing construction of the 50MW(e) and 200MW(e) reactors, foregoing reprocessing, ceasing activities at the reprocessing plant and ceasing operation of the fuel fabrication plant. North Korea will not construct any new graphite-moderated reactors or related facilities.

Stage Two: Implementation of Full-scope Safeguards

When a 'significant portion' of the LWR project is completed, but before delivery of key nuclear components, North Korea will come into full compliance with its safeguards agreement, including permitting the IAEA access to additional sites and information deemed necessary by the IAEA to verify the accuracy and completeness of North Korea's initial report on all nuclear material in North Korea.

Stage Three: Removal of Spent Fuel

When delivery of key nuclear components for the first LWR unit begins, the transfer of 5MW(e) spent fuel from North Korea will begin, and transfer of the fuel will be completed when the first LWR unit is completed. In the meantime, following discussions with the US, North Korea will select and begin to implement a method of spent fuel storage that permits transfer of the fuel.

Stage Four: Dismantlement of Facilities

When the first LWR unit is completed, North Korea will begin dismantling its graphite-moderated reactors and related facilities, and the dismantlement will be completed when the second unit of the LWR project is completed.

Source: The Agreed Framework and Confidential Minute

'significant portion' of the LWR project was completed, rather than run the risk that the Agreed Framework would collapse if North Korea refused to cooperate with the IAEA in accounting for its pre-1992 plutonium production. Critics feared that the momentum of the LWR project would increase pressure on the IAEA to give North Korea a clean bill of health in order to keep the project alive and avoid a costly interruption. Some US officials believed the North Korean regime was likely to collapse before the LWR project could ever be completed, whereas critics argued that the assistance provided to North Korea under the Agreed Framework would help to prop up the regime.

Implementation of the Agreed Framework (1994–97)
The Agreed Framework required many moving parts, and implementation proved to be complicated and often difficult. Within a few months, the IAEA installed equipment and established a year-round inspector presence at Yongbyon to verify that North Korea's plutonium production facilities were frozen, while Washington quickly arranged for the initial delivery of 50,000 tonnes of heavy fuel oil. By January 1995, American and North Korean technical experts reached agreement to store the 8,000 spent fuel rods in stainless steel cans at the Yongbyon spent fuel pond, although it would take five years to complete the canning process due to various technical difficulties. In March 1995, the US, South Korea, and Japan formed an international consortium called the Korean Peninsula Energy Development Organization (KEDO) to provide annual deliveries of 500,000 tons of HFO to North Korea and implement the multi-billion dollar LWR project. To help pay for HFO deliveries, which turned out to be more expensive than anticipated, the US had to gather contributions from Australia, Canada, Finland, New Zealand, Indonesia, Chile, Argentina and others. In September 1997, the European Union (EU) joined as a voting member of the Executive Board in exchange for a substantial multiple-year contribution.

Working out the details of the LWR project required a series of difficult negotiations with North Korea. These became enmeshed in North–South tensions as Pyongyang sought to substitute a reactor from another country instead of South Korea. In the end, North Korea accepted the 'South Korean-type' reactor, and an LWR Supply Agreement between North Korea and KEDO specifying the details of the project was completed in December 1995. Even then, implementation of the LWR project moved only slowly, with disagreements over such issues as wages for

North Korean labourers working on the project and the communications and transportation links between the construction site (at Sinpo) and South Korea. In late 1996, just as the LWR project was gathering steam, an incident involving a North Korean miniature submarine, which foundered off South Korea's coast and was presumed to have been involved in espionage activities, abruptly halted the project and set it back by months. Because of these delays, the project fell behind in a schedule that envisaged completion by the 'target date' of 2003, and Pyongyang began to demand 'compensation' for the delay.

Aside from the delays in implementing the LWR project, difficulties emerged in carrying out the Agreed Framework's political provisions. Pyongyang complained that Washington did not fulfill the Agreed Framework's provisions for reducing economic sanctions, and negotiations to establish diplomatic liaison offices in Pyongyang and Washington broke down over the issue of whether the US could transit the Demilitarised Zone (DMZ) separating North and South Korea with diplomatic pouches. In April 1996, Washington launched a proposal for Four Party Talks – between the US, South Korea, North Korea and China – to discuss proposals for introducing confidence-building measures on the Peninsula and establishing a formal peace treaty to replace the 1953 armistice in place since the Korean War. But little progress was made in several rounds of talks. North–South relations also remained prickly. In 1997, at the end of his term, South Korean President Kim Young Sam made a new effort to engineer a North–South summit, but he was rebuffed by Pyongyang. Relations between North and South Korea did not turn for the better until the December 1997 election of President Kim Dae Jung, who instituted a 'sunshine policy', featuring heavy emphasis on inducements, to improve relations with Pyongyang.

Another difficult issue at this time concerned US efforts to restrain North Korean missile exports and missile development. During the Agreed Framework negotiations, Ambassador Gallucci had warned Kang Sok Ju that North Korean missile exports to Iran would damage prospects for improving ties with Washington, and North Korea apparently responded by delaying an impending shipment of *No-dong* missiles to Iran. By 1995, however, *No-dong* exports to Iran had begun, and the US pressed North Korea to begin missile talks. The first round was held in April 1996, followed by a second round in June 1997. In these talks, the US sought limits on both North Korean missile exports and indigenous development. North Korea refused to

'From Washington's perspective, the Agreed Framework's political and economic provisions were designed to build up incentives for Pyongyang to sacrifice its residual nuclear weapons capability'

discuss its indigenous missile programme, which it said was linked to broader security issues on the Peninsula, including the presence of US forces. At the same time, North Korea said it would end exports in exchange for cash 'compensation' for lost revenues. While ruling out cash, the US side offered to take additional steps to improve bilateral relations and lift additional economic sanctions if North Korea ended missile exports.

The Agreed Framework in crisis (1998–2000)

In Washington, criticism of the Agreed Framework became involved in partisan politics. After the Republican Party gained control of Congress in the November 1994 mid-term elections, the Clinton administration was unable to secure strong political support for the agreement, resulting in annual battles over funding for KEDO. Congress also imposed requirements for various Presidential certifications as a condition for the US providing funding to KEDO for HFO deliveries. Congressional criticism of the Agreed Framework peaked in summer 1998 because of two developments. Firstly, in early August, the *New York Times* broke a story that North Korea was building a vast underground facility at Kumchang-ri, which US intelligence believed was intended to house a secret plutonium production reactor and reprocessing facility. Evidence for this assessment was too circumstantial and premature for a firm conclusion because construction was still at an early stage. Nonetheless, it was credible to US analysts and officials that North Korea would seek to develop an alternative source of fissile material in order to maintain its nuclear weapons capability even after the Agreed Framework required North Korea to declare any plutonium it had produced before 1992. Congress had been briefed about the Kumchang-ri site prior to the press leak, but the publicity provided political ammunition to critics of the Agreed Framework and increased pressure on the administration to prove in definite terms that North Korea was complying with the deal.

Secondly, North Korea's launch of a *Taepo-dong* missile on 31 August 1998 (which overflew Japan) highlighted concerns that North Korea was attempting to develop long-range missiles that could threaten the US. Through satellite intelligence, Washington had detected preparations for the *Taepo-dong* launch, and strongly warned Pyongyang not to proceed, but to no avail. Pyongyang asserted that the *Taepo-dong* launch was a satellite launch-attempt in celebration of Kim Jong Il's ascension to power. In Washington, the *Taepo-dong*

launch, coming on the heels of the Kumchang-ri allegations, provoked strong Congressional opposition to the administration's North Korea policy and threatened to cut off US funding for KEDO. In addition, the launch increased political pressure on the administration to proceed with rapid development and deployment of a national missile defense shield, even if it required withdrawal from the 1972 Anti-Ballistic Missile (ABM) Treaty, a move opposed by the White House.

In response to the Kumchang-ri revelations and the *Taepo-dong* launch, Washington took several steps. Firstly, the US began negotiations with North Korea to obtain access to the Kumchang-ri site in order to determine whether it was actually a secret nuclear facility. In March 1999, after several rounds of talks, North Korea agreed to a US 'visit' to the underground site in exchange for additional US food assistance to North Korea. When the US team visited the site in May 1999, they found a vast series of underground tunnels and rooms, although the dimensions and configuration of the underground site could not house a reactor and reprocessing facility. The real purpose of the site remains unknown. Secondly, in response to the *Taepo-dong* launch, Washington warned Pyongyang that additional missile tests would jeopardise US support for the Agreed Framework and humanitarian food shipments. Going a step further, Japan temporarily suspended funding for KEDO and the LWR project. In September 1999, following several rounds of negotiations between North Korean and US diplomats, Pyongyang agreed to a moratorium on additional long-range missile tests, which covered both the *No-dong* and *Taepo-dong* missiles, in exchange for the US lifting a number of economic sanctions. North Korea's acceptance of the moratorium may have been encouraged by pressure from China and Russia, who feared that additional missile tests would further stimulate US efforts to develop missile defences, which they opposed.

Finally, in November 1998, the White House asked former Secretary of Defense William Perry to conduct a comprehensive review of US policy towards North Korea and make recommendations for improvements, fulfilling a Congressional requirement for such a review. In October 1999, Secretary Perry released his report, recommending that the US offer to normalise relations with North Korea and lift economic sanctions if North Korea agreed to freeze and eventually dismantle its long-range missile force and end missile exports. In essence, the 'Perry Report' proposed an Agreed Framework for missiles, with some additional measures to deal with North Korea's nuclear programme. If the

'It was credible to US analysts and officials that North Korea would seek to develop an alternative source of fissile material in order to maintain its nuclear weapons capability'

North did not accept that path, the report recommended that the United States and its allies should take other steps 'to ensure their security and contain the threat.'

The almost missile deal

After the release of the 'Perry Report', Washington urged Pyongyang to begin serious negotiations on resolving the missile issue while the Clinton administration was still in office, but Pyongyang bided its time and focused energies on other efforts, including the historic summit in Pyongyang between Chairman Kim Jong Il and President Kim Dae Jung in June 2000. Little progress was made in another round of missile talks in July 2000. In the talks, US negotiators suggested that other countries could provide satellite launch services to North Korea if it agreed to forgo further development of long-range missiles, but Pyongyang continued to link its indigenous missile programme to broader security issues. Concerning missile exports, North Korean negotiators maintained their requirement for cash compensation, but began to hint that other forms of compensation might be considered.

Towards the end of the Clinton administration, North Korea proposed a grand bargain on missiles. In July 2000, during a visit to Pyongyang by Russian President Vladimir Putin, Kim Jong Il floated a proposal for a comprehensive missile deal that would include guaranteed North Korean access to international launches for its satellites in exchange for limits on North Korea's missile development. Initially, Washington was sceptical of the North Korean proposal, but further details emerged during the visit of North Korean Vice Marshall Cho Myong Rok to Washington in September 2000. This was followed one month later by a visit to Pyongyang by Secretary of State Madeleine Albright who travelled in pursuit of a potential missile deal. According to Chairman Kim Jong Il, North Korea would freeze the development, production, deployment, and testing of missiles over 500km range if the US guaranteed that other countries would launch a few North Korean civilian satellites every year at no cost. In addition, North Korea proposed to end all missile and missile-related exports in exchange for compensation in unspecified goods rather than cash. From Pyongyang's standpoint, the missile deal was linked to broader steps to improve US–North Korea relations, including a visit by President Clinton to Pyongyang and establishment of diplomatic relations.

Before agreeing to any visit by President Clinton, however, Washington sought to clarify the North Korean proposal. Missile experts from the two sides met in Malaysia in November 2000. During these discussions, agreement was reached on several key issues. North Korean negotiators agreed that the ban on missile exports would be comprehensive, including missiles themselves, missile components, materials, equipment, and technology. North Korea also agreed that any satellites it provided for launch under the agreement would be purely civilian and that safeguards would be instituted to ensure that North Korea would not obtain access to missile technology in connection with launches of its satellites by other countries.

But a number of key issues remained unresolved. Firstly, the type of missile covered by the freeze was unclear. North Korean diplomats agreed that their proposal banned further development and production of No-dong and Taepo-dong-type missiles, but they were not prepared to include possible Scud variants with a range over 500km. The US proposed a more restrictive threshold defined by the Missile Technology Control Regime – namely, missiles capable of delivering a warhead of 500kg to 300km – but North Korean negotiators argued that limits on its Scud forces could only be considered in the context of broader security issues on the Korean Peninsula. In essence, North Korea sought to exempt its Scud missile force from the freeze.

Secondly, the disposition of existing missiles covered by the freeze was not resolved. The US proposed that North Korea agree to eliminate its existing No-dong and Taepo-dong missiles and production facilities, but North Korean negotiators said they had no instructions on this issue. Privately, however, the North Koreans hinted that they could consider gradual elimination of No-dong and Taepo-dong missiles over an extended period of time in exchange for unspecified compensation. Thirdly, both sides recognised that verification and monitoring procedures would need to be specified in detail. North Korean officials agreed to the concept that some 'cooperative' measures would be necessary to verify a missile agreement, yet they strongly opposed on-site 'inspections'. But again, in private, North Korean officials hinted that they might accept 'visits' to missile facilities in the context of converting these facilities to civilian use. Finally, the type and size of the compensation package to end missile exports was not agreed. Although North Korean officials suggested that food or oil would be welcome, there was no agreement on the amount and nature of compensation.

Although none of these outstanding issues appeared insurmountable, the Clinton administration ran out of negotiating time. North Korea promised that all issues

'According to Chairman Kim Jong Il, North Korea would freeze the development, production, deployment, and testing of missiles over 500km range'

could be resolved once the two presidents sat down together in Pyongyang, but the White House was not willing to risk such a controversial visit without prior agreement on key issues. This tactical standoff, combined with the delayed outcome of the US presidential elections and President Clinton's focus on Middle Eastern peace negotiations in his final months in office, doomed the effort to complete a US–North Korea missile deal.

A broad agenda and bold approach

The administration of President George W. Bush that took office in January 2001 had divergent views on North Korea. Many incoming officials saw the Agreed Framework as paying blackmail to prop up a rogue regime that could not be trusted to honour its commitments. Rather than reward Pyongyang for its bad behaviour, these officials argued, the US should adopt a strategy of containment and isolation, hoping to hasten the collapse of the North Korean regime and remove the problem at its roots. Other officials, however, argued that the Agreed Framework had succeeded in freezing North Korean plutonium production since 1994, and they supported further diplomatic efforts to limit North Korea's nuclear and missile programmes. These divergent views were on display during the March 2001 visit of South Korean President Kim Dae Jung, who sought Washington's endorsement for his 'sunshine policy' of engagement with North Korea. While Secretary of State Colin Powell initially indicated that Washington intended to continue the missile talks begun by the Clinton administration, President Bush publicly expressed scepticism about the value of engagement with North Korea in a difficult meeting with President Kim Dae Jung. Subsequently, Secretary Powell clarified that any resumption of negotiations with Pyongyang would require completion of an overall review of US policy towards North Korea.

On 6 June 2001, the Bush administration issued a policy statement on North Korea that reflected a compromise among different views. On the one hand, the US committed itself to continuing support for the Agreed Framework – including funding for heavy fuel oil (HFO) deliveries – as long as North Korea upheld its end of the bargain. In fact, the Bush administration obtained Congressional support for a significant expansion in funding for HFO to $90 million in fiscal year 2002. On the other hand, Washington stated that any future negotiations should pursue a 'broad agenda', including 'improved implementation of the Agreed Framework relating to North Korea's nuclear activities;

verifiable constraints on North Korea's missile programs and a ban on its missile exports; and a less threatening conventional military posture.' In return for North Korean actions on the 'broad agenda', Washington said that it would 'expand our efforts to help the North Korean people, ease sanctions, and take other political steps.' In the meantime, the US would continue to provide humanitarian food assistance to North Korea, and offered to resume discussions 'without condition.'

In response to Washington's decision not to resume negotiations for a stand-alone missile deal, Pyongyang rebuffed repeated US offers to hold bilateral talks and tried to solicit international support for its proposed missile agreement. In May, Kim Jong Il told a visiting delegation from the EU that North Korea would extend its missile test moratorium until at least 2003, and even suggested that Europe might play a role in negotiations to limit North Korea's missile programme. In August, Kim Jong Il travelled to Moscow for meetings with President Putin, seeking to enlist his support to revive missile talks. Washington, however, continued to insist on a broad agenda as a basis for resuming bilateral talks – a stance which critics feared would paralyse negotiations.

The terrorist attacks of 11 September 2001 further complicated prospects for US–North Korea negotiations. In Washington, the attacks galvanised fears of a new threat posed by the combination of international terrorism and the proliferation of weapons of mass destruction (WMD), focused in the first instance on 'rogue regimes' which support terrorism and pursue WMD. Even though North Korea has not been associated with terrorism for many years, it was included in the 'axis of evil' pronounced by President Bush in his State of the Union address of January 2002. Pyongyang, already suspicious of Washington's intentions, declared that the speech demonstrated that Washington was determined to 'stifle' its regime. Even after the President's speech, however, Secretary Powell stressed that the US was still prepared to resume discussions with Pyongyang. The 'axis of evil' speech also created alarm in South Korea, which feared that US hostility towards North Korea would increase tensions on the Peninsula and sour North–South relations. Seeking to maintain solidarity with Seoul, President Bush avoided criticism of North Korea and, during his visit to South Korea in late February 2002, said that the US had no intention of invading North Korea.

In late April 2002, North Korea finally agreed to resume bilateral discussions. In preparation for the

---

'The administration of President George W. Bush that took office in January 2001 had divergent views on North Korea' ... *'The terrorist attacks of 11 September 2001 further complicated prospects for US–North Korea negotiations'*

talks, Washington prepared a 'bold approach' to offer North Korea substantial economic and political benefits if it completely gave up WMD and missiles, withdrew conventional forces from near the demilitarised zone with South Korea, and improved human rights. Even more than the earlier 'broad agenda', the 'bold approach' was intended to test Pyongyang's willingness to accept a quick, comprehensive resolution of outstanding issues, although many in Washington were sceptical that North Korea would agree to such a dramatic proposal. In any event, following a clash between North and South Korean naval forces in late June, Washington cancelled a trip to Pyongyang, planned for Assistant Secretary of State for East Asian and Pacific Affairs James Kelly on 10 July, to present this new proposal. By late July, Pyongyang had expressed regret over the loss of life in the naval incident and sought to reschedule the meeting with the US.

The enrichment bombshell
Around the time that Washington cancelled the Kelly visit, the US intelligence community issued a secret assessment concluding that North Korea had embarked on a clandestine programme to produce weapons-grade uranium with gas centrifuge technology that it had obtained from Pakistan in exchange for *No-dong* missiles. Although little was known about the programme, the CIA provisionally assessed that North Korea was constructing a plant that could produce enough weapons-grade uranium for two or more nuclear weapons per year when fully operational, which could be by 'mid-decade'. Initially, Washington was divided on how to react to the discovery that North Korea was in violation of the Agreed Framework and the NSDD and most likely the NPT as well. However, after the surprisingly successful summit between Japanese Prime Minister Junichiro Koizumi and Chairman Kim Jong Il in Pyongyang on 17 September, the US felt compelled to confront North Korea before Tokyo concluded any agreement with Pyongyang to normalise relations and provide economic assistance.

On 4–5 October 2002, Assistant Secretary Kelly travelled to Pyongyang to meet Vice Foreign Minister Kang Sok Ju and other North Korean officials. In the initial meeting, Kelly outlined Washington's 'bold approach', but said that the US could not improve relations until North Korea dismantled its clandestine uranium enrichment programme. According to American accounts, North Korean officials initially denied the accusations and reminded the US of the

earlier Kumchang-ri incident, in which Washington had falsely accused North Korea of building a secret underground reactor. But in the following day of meetings, US officials said later, Kang angrily acknowledged the enrichment programme, and he said it was justified by the Bush administration's threats and hostility. Kelly pointed out that North Korea had begun the enrichment programme several years before President Bush took office.

North Korea's 'acknowledgement' that it was pursuing a secret enrichment programme took Washington by surprise. Some US officials saw it as another example of 'confession diplomacy', like Kim Jong Il's personal admission to Prime Minster Koizumi that North Korea had abducted Japanese citizens in the 1970s and 1980s. For others, it was a brazen act of North Korean defiance. With many US officials already sceptical about the wisdom and morality of the Agreed Framework, North Korea's admission strengthened the case for renouncing the Agreed Framework and resisting North Korean 'blackmail'. Washington was determined not to 'reward' Pyongyang's actions by offering fresh incentives to abandon its enrichment programme, which was already banned under existing agreements. At the same time, with its energies fully focused on the mounting diplomatic and potential military campaign against Iraq, Washington had little enthusiasm for a confrontation with Pyongyang that would divert attention away from Iraq and complicate relations with South Korea and Japan, who were wary of pushing North Korea into desperate actions.

As a result, Washington's initial reactions were cautious: to pressure Pyongyang but not provoke a crisis. On 16 October, Washington announced that North Korea had 'acknowledged' that it was pursuing a clandestine enrichment programme in violation of the Agreed Framework and other agreements and called on the North to 'eliminate its nuclear weapons program in a verifiable manner'. At the same time, Washington emphasised that it sought a 'peaceful resolution of the situation' in close consultation with South Korea and Japan, and held out the prospect of discussing economic and political measures to 'improve the lives of the North Korean people' if the North complied with its nuclear obligations. Washington demanded that North Korea abandon its nuclear weapons programme as a basis for any further bilateral discussions on improved relations.

North Korea responded to the US announcement with a Foreign Ministry statement on 25 October 2002. Pyongyang complained that the US had produced 'no

'The CIA provisionally assessed that North Korea was constructing a plant that could produce enough weapons-grade uranium for two or more nuclear weapons per year' … 'Washington was determined not to 'reward' Pyongyang's actions by offering fresh incentives to abandon its enrichment programme, which was already banned under existing agreements'

evidence' that it was breaking the Agreed Framework and instead accused the US of violating the deal by failing to deliver the LWR project on time and provide formal assurances against the threat or use of nuclear weapons. Pyongyang rejected Washington's proposal that it disarm as a condition for talks, but offered to negotiate a settlement on three conditions: 'Firstly, if the U.S. recognizes the DPRK's sovereignty, secondly, if it assures the DPRK of non-aggression and thirdly, if the U.S. does not hinder the economic development of the DPRK.'

Just as it had nearly a decade earlier when North Korea first threatened to withdraw from the NPT, the US sought to mobilise international political and economic pressure against North Korea. In the meantime, however, Seoul's position had fundamentally changed. Anxious to preserve his 'sunshine policy', President Kim Dae Jung refused to make South Korean humanitarian assistance and economic cooperation dependent on North Korea dismantling its nuclear weapons programme. This issue, with strong anti-US overtones, became enmeshed in the South Korean presidential elections. Tokyo was more willing to support US efforts, especially since North Korea's admissions over the abduction issue had inflamed Japanese public opinion. In late October, Japan informed North Korea that normalisation would depend on resolution of the nuclear issue, which quickly led to a breakdown in normalisation talks between the two nations.

The most immediate issue confronting the allies was what to do about KEDO. With North Korea in violation of the Agreed Framework, Washington pressed for a 'suspension' of KEDO activities, starting with HFO shipments (which were largely financed by America). Seoul and Tokyo, however, were concerned that Pyongyang would retaliate by resuming nuclear activities frozen under the Agreed Framework, and argued that KEDO should continue oil shipments until its current funds ran out in January 2003. In a compromise, the KEDO Executive Board announced on 14 November that the November oil shipment – then en route to North Korea – would be delivered, but that additional shipments would be suspended, starting in December. To reassure Seoul and Tokyo, the White House issued a statement on 15 November, welcoming KEDO's decision and reiterating that 'the United States has no intention of invading North Korea'. Instead, Washington said North Korea could 'benefit from participation in the international community' if it 'completely and visibly' eliminated its nuclear weapons programme.

Death of the Agreed Framework

At the time of the KEDO decision, Washington expressed confidence that North Korea was too weak and isolated to retaliate for the suspension of HFO supplies, and that it would be forced by international pressure and the threat of sanctions to dismantle its nuclear weapons programme. Pyongyang's initial response seemed to bear out Washington's prediction. In an official statement on 21 November, Pyongyang repeated accusations that Washington was violating the Agreed Framework, mixing these reproaches with renewed offers to negotiate a solution of the nuclear issue based on a 'non-aggression' pact with the US. On 12 December, however, just as it seemed that the danger of retaliation had passed, Pyongyang announced that it was restarting its 5MW(e) reactor and resuming construction of the larger 50MW(e) and 200MW(e) reactors. At the same time Pyongyang also suggested it would consider a 'refreeze' of the reactors 'depending on the attitude of the U.S.'. In the meantime, North Korea began to load fresh fuel into the 5MW(e) reactor, in view of IAEA inspectors at the site.

Although Washington did not anticipate North Korea's decision to 'unfreeze' its graphite-moderated reactors, the announcement was seen as a relatively cautious move – intended to increase political pressure on Washington, rather than precipitate a full-blown crisis. As a practical matter, restarting the 5MW(e) reactor did not present an immediate threat because the facility could not produce a significant amount of additional plutonium for at least a year, and the larger 50MW(e) and 200MW(e) reactors could not be completed for several years at best. While calling North Korea's decision 'unacceptable', Washington took no immediate action, as it awaited the outcome of South Korea's presidential elections between the Millennium Democratic Party candidate, Roh Moo Hyun, who supported Kim Dae Jung's Sunshine Policy, and Grand National Party candidate, Lee Hoi Chang, who advocated a tougher policy towards Pyongyang. From Washington's perspective, the election of Lee would strengthen its hand against North Korea, but the narrow victory of Roh on 19 December, partly on a wave of anti-American feeling, may have given Pyongyang a sense that it was in a stronger position to exploit differences between Washington and Seoul and play more of its nuclear cards.

On 22 December 2002, North Korea moved to completely unfreeze its plutonium production facilities by ordering the IAEA to remove surveillance cameras and seals on the 5MW(e) reactor, the spent fuel storage

'With North Korea in violation of the Agreed Framework, Washington pressed for a 'suspension' of KEDO activities'

pond and the reprocessing facility. This was followed, on 27 December, by the expulsion of the inspectors themselves. The removal of equipment and inspectors denied any real-time capability to monitor North Korean nuclear activities at Yongbyon. More significantly, North Korea also announced on 27 December that it would soon complete preparations to resume operations at the reprocessing facility. This, according to Pyongyang, was a safety step necessary to handle spent fuel from its newly unfrozen reactors. The implicit threat, however, was to extract the estimated 25–30kg of plutonium contained in the 8,000 spent fuel rods removed from the 5MW(e) reactor in 1994 – enough for a few nuclear weapons.

By the end of December, the Agreed Framework was essentially dead. Washington, however, remained determined not to let North Korean brinksmanship distract it from dealing with Iraq or force it to negotiate under North Korean duress. In response to the end of the freeze, the US and its allies agreed on a cautiously worded resolution passed by the IAEA Board of Governors on 6 January 2003. This called on North Korea to allow the return of inspectors and restoration of monitoring equipment, and hinted that the IAEA would otherwise report North Korean non-compliance to the UNSC. The resolution was advertised as a 'last chance' for North Korea to restore the freeze. At the same time, Washington offered a small diplomatic concession in a trilateral US–South Korea–Japan statement issued on 7 January. In the statement, Washington eased its previous refusal to meet with North Korea until Pyongyang abandoned its nuclear weapons programme, saying 'the United States is willing to talk to North Korea about how it will meet its obligations to the international community'. As US officials explained, this meant that Washington was willing to meet with Pyongyang to hear how it intended to dismantle its nuclear weapons programme.

North Korea, however, did not find this offer attractive. Perhaps encouraged by Washington's focus on Iraq, Pyongyang escalated further. In response to the IAEA Board of Governors resolution, Pyongyang formally withdrew from the NPT on 10 January, to free itself, North Korea explained, from any safeguards obligations. Although the Treaty requires a 90-day notice before withdrawal can take effect, Pyongyang argued that it had already given the required notice in March 1993, when it had originally declared its intent to withdraw. In an effort to minimise international condemnation, Pyongyang offered the reassurance that 'Though we pull out of the Treaty, we have no intention

to produce nuclear weapons and our nuclear activities at this stage will be confined only to peaceful purposes such as production of electricity'. Echoing its statement of ten years earlier, Pyongyang blamed its decision on the actions of the US and IAEA and offered to demonstrate that its nuclear programme was peaceful 'if the US drops its hostile policy to stifle the DPRK and stops its nuclear threat to the DPRK'. Unlike 1993, however, North Korea proposed that its nuclear facilities be inspected by US experts, rather than the IAEA.

After North Korea's withdrawal from the NPT, Washington sought to contain further tit-for-tat escalation. As its troops began to mobilise in the Persian Gulf for *Operation Iraqi Freedom*, the US could not afford another crisis. In order to contain the confrontation with North Korea, the US proposed to begin negotiations in a multilateral rather than bilateral context. From Washington's perspective, multilateral talks would avoid any appearance that North Korea's 'blackmail' had forced Washington into direct negotiations, and the US hoped to enlist the involvement of additional parties to increase pressure on North Korea to dismantle its nuclear programme and honour any new agreement. In late January, the US privately offered to meet North Korea in '5 plus 5' multilateral talks involving the permanent UNSC members (the US, Russia, China, the UK and France) plus South Korea, Japan, North Korea, Australia and the EU. But Pyongyang insisted it would only accept direct bilateral negotiations with Washington. For the same reason that Washington found multilateral talks attractive, Pyongyang feared that any additional participants would naturally side with the US against North Korea.

Tensions increased in February. On 5 February, North Korea announced that it was 'now putting the operation of its nuclear facilities for the production of electricity on a normal footing after their restart', apparently referring to the start-up of the 5MW(e) reactor. In addition, satellite imagery detected increased heavy vehicle activity at the Yongbyon spent fuel storage facility, presumably moving some of the 8,000 spent fuel rods to the reprocessing facility or to another storage facility less vulnerable to military attack. On 12 February, the IAEA Board of Governors convened a special meeting and formally found North Korea in violation of its NPT safeguards obligations and reported the matter to the UNSC, in theory setting the stage for action by the Security Council (although permanent members China and Russia continued to regard this as provocative). In response to indications that North Korea might be

'Perhaps encouraged by Washington's focus on Iraq, Pyongyang escalated further '... 'In order to contain the confrontation with North Korea, the US proposed to begin negotiations in a multilateral rather than bilateral context'

planning to start reprocessing, Washington decided to deploy additional bombers and stealth aircraft to the region. Although Washington had not decided to risk a pre-emptive strike on the North Korean reprocessing facility, the move was seen as useful psychological warfare against Pyongyang, which believed that such a strike was under consideration. When press reports appeared in early February that the US was putting long-range bombers on alert for possible deployment in the Korean theatre, Pyongyang publicly warned that it might launch a pre-emptive strike of its own if the US built up threatening forces. In a dangerous incident on 1 March 2003, four North Korean fighters harassed an American RC-135 reconnaissance plane over international waters. In early March, in connection with the *Foal Eagle* US-South Korea military exercise, the US deployed 24 long-range bombers to Guam and a small number of F-117A stealth aircraft to South Korea.

China steps in
In the midst of these rising tensions, Secretary Powell visited Beijing in late February and suggested that China seek to persuade North Korea to agree to five party talks involving the US, China, Japan, and North and South Korea. For Beijing, the stalemate and rising tensions presented both a threat and an opportunity. Initially, Beijing had reacted to the breakdown of the Agreed Framework as it had to North Korea's threat to withdraw from the NPT in 1993. Publicly, Beijing advocated support for a nuclear-free Korean Peninsula achieved through peaceful dialogue; privately, Chinese officials minimised their influence with Pyongyang and urged the US to solve the dispute directly with North Korea. With tensions rising in early 2003, however, China apparently decided that it needed to play a more active role to avoid the risks of instability and confrontation on its doorstep. At all costs, it wanted to avoid a situation in which North Korea's regime might collapse under US and other pressure (leaving China to bear many of the economic and political consequences of the resulting vacuum) or in which a general war on the Peninsula might be provoked. Additionally, Beijing feared that tensions on the Peninsula – and provocative nuclear actions by Pyongyang – might provide a strong rationale for Japan to substantially develop its defence capabilities, possibly including the development of a Japanese nuclear deterrent. At the same time, China may have seen an opportunity to improve its diplomatic position in the region and strengthen relations with the US, which were improving after the 11 September terrorist attacks. From Beijng's perspective, the more indispensable China

became to resolving the North Korea threat, the more it could influence US policy on Taiwan, the primary foreign policy issue on China's agenda.

In early March, Chinese Vice Premier Qian Qichen travelled to Pyongyang for talks with Kim Jong Il, and proposed a compromise three party (US–China–North Korea) formula as an alternative to the US proposal for five party talks, suggesting that direct talks between the US and North Korea could take place on the margins of the meeting hosted in Beijing. Reportedly, Chinese oil supplies to North Korea were temporarily interrupted for 'technical reasons' to encourage Pyongyang to accept the Chinese compromise. At the same time, to avoid antagonising North Korea, China, along with Russia, blocked any action by the Security Council on 9 April. On 12 April, Pyongyang announced that it would not 'stick to any particular dialogue format' for negotiations to resolve the nuclear issue, signalling that it was prepared to accept the Chinese proposal. To reinforce its bargaining leverage before the talks, North Korea announced on 18 April that 'we are successfully reprocessing more than 8,000 spent fuel rods at the final phase' although US intelligence did not detect any indications that reprocessing had begun. While Washington was reluctant to drop the participation of South Korea and Japan, it felt it could not afford to antagonise Beijing by rejecting its compromise offer and, after consulting with Seoul and Tokyo, who both concurred, Washington agreed.

The Three Party Talks, held in Beijing on 24–25 April 2003 went badly. Pyongyang proposed a new agreement, in which the US and North Korea would take simultaneous steps in a series of stages, eventually leading to the dismantlement of North Korea's nuclear weapons programme. In the first stage, North Korea would declare its intent to abandon its nuclear weapons programme, while KEDO resumed oil shipments. In the second stage, North Korea would allow resumed inspections of its nuclear facilities, and Washington would sign a non-aggression pact with Pyongyang. In the third stage, missile issues would be dealt with once political relations between Pyongyang and Washington and Tokyo were normalised. Finally, and only once the light water reactor project was completed, North Korea would dismantle its nuclear capability.

For Washington, the North Korean proposal was totally unacceptable, and the US delegation, headed by Assistant Secretary Kelly, reiterated the US position that North Korea must disarm 'completely irreversibly, and verifiably' before receiving any political or economic benefits. Aside from these substantive

'With tensions rising in early 2003, China apparently decided that it needed to play a more active role to avoid the risks of instability and confrontation on its doorstep' ... 'The Three Party Talks, held in Beijing on 24–25 April 2003 went badly'

differences, the US delegation was under strict instructions not to meet bilaterally with the North Korean side, while the North Koreans came to Beijing expecting a formal bilateral meeting with the US. In the end, the head of the North Korean delegation, Li Gun (a relatively junior official) had to find an informal opportunity to warn Kelly that North Korea already had nuclear weapons, had completed reprocessing the fuel rods and might take additional actions if no agreement was reached. The talks ended one day early.

After the breakdown of the April meeting, Washington and Pyongyang hardened their positions. In early May, the US decided that any future talks should be expanded to include South Korea and Japan, and the administration began to develop plans to strengthen international efforts to interdict shipments of proliferation-related goods, primarily directed against North Korea, which resulted in President Bush's announcement of the Proliferation Security Initiative on 31 May 2003. For its part, Pyongyang 'nullified' the North-South Denuclearization Declaration on 12 May, just as South Korean President Roh was coming to Washington for his initial meeting with President Bush. More ominously, US intelligence detected some indications in May and June that North Korea had begun reprocessing the spent fuel rods, suggesting that North Korea might be carrying out the private warnings it had conveyed to US diplomats on the eve of the Three Party Talks in April.

In July 2003, North Korean diplomats told American officials privately that they had completed reprocessing the 8,000 fuel rods, although US intelligence could not independently verify this claim. Most US analysts thought that North Korea probably carried out some limited reprocessing operations at the Yongbyon facility in May and June, perhaps enough for a weapon or two, but had stopped short of reprocessing the entire 8,000 rods. Perhaps, observers speculated, North Korea ran into technical difficulties, or had limited reprocessing in response to strong private warnings from Washington and Beijing that reprocessing would scuttle chances for negotiations. Alternatively, US detection capabilities were too weak to confirm that North Korea had indeed completed reprocessing the spent fuel on hand.

Behind the scenes, China sought to find a formula for multilateral talks, while continuing to protect Pyongyang from international pressure. In early July, China and Russia again blocked action by the UN Security Council, arguing that any statement from the Council criticising Pyongyang could disrupt delicate efforts to resume multilateral talks. At the end of July, after working through several different formulations, China finally orchestrated a compromise. Korea agreed to accept Six Party Talks (involving the US, South Korea, North Korea, China, Japan and Russia), and the US agreed to accept a bilateral US–North Korea meeting on the margins of the multilateral meeting. Reportedly, Beijing helped to secure Pyongyang's acceptance of the compromise with the inducement of extra food and oil deliveries.

In the first round of the Six Party Talks, held in Beijing on 27–29 August 2003, the heads of both the US and North Korean delegations, Assistant Secretary Kelly and Deputy Foreign Minister Kim Yong Il, held a very short bilateral meeting, but the wider talks produced little progress on substance. North Korea reiterated its April proposal for simultaneous steps eventually leading to disarmament, and hinted that it could accept a freeze on its nuclear activities as a first step towards disarmament. At same time, North Korea denied US accusations that it had an enrichment programme and reportedly threatened to declare its nuclear weapons status and conduct a nuclear test if no solution was reached. Privately, North Korean officials explained that Kelly had 'misunderstood' Kang's statements in October 2002. They also expressed disappointment that the US did not suggest a counter-offer to North Korea's April 2003 proposal, but US officials maintained that their presentation included the concept that North Korean disarmament could take place in a series of stages, therefore signalling that the US might be prepared to provide some inducements even before 'complete, irreversible, and verifiable' disarmament was achieved.

The first round of Six Party Talks ended without being able to reach agreement on a joint communiqué, but Beijing issued a chairman's statement, which it said reflected general principles that all the parties agreed for resolving the nuclear dispute. Among these, Beijing said, the parties agreed on a peaceful settlement to achieve a nuclear-free Korean Peninsula that addresses North Korea's security concerns through 'stages and through synchronous or parallel implementation in a just and reasonable manner'. China also announced that all parties agreed to hold another round of Six Party Talks as soon as possible. Hoping to schedule these for mid-October, China sought to mediate agreement on a draft 'statement of principles' that could be announced at the conclusion of the next round of talks. The object of this statement would be to create an overall political framework for the talks and build confidence and a sense of continuity.

'On 2 October 2003, North Korea publicly announced it had successfully finished reprocessing the spent fuel rods and was using the resulting plutonium to increase its "nuclear deterrent force"'

However, sensing that Washington was eager to resume talks, North Korea declared that it saw no value in any more Six Party Talks, perhaps playing hardball to raise the price for agreeing to another round. On 2 October 2003, North Korea publicly announced it had successfully finished reprocessing the spent fuel rods and was using the resulting plutonium to increase its 'nuclear deterrent force'. The US was still uncertain whether North Korea's claim was accurate and, in the absence of inspectors on the ground, had no way of determining whether any separated plutonium had been fabricated into nuclear weapons. Nonetheless, some diplomatic progress was made. On 19 October, in a meeting with Chinese President Hu Jintao on the margins of an Asia–Pacific Economic Cooperation (APEC) summit in Bangkok, President Bush indicated that the US was prepared to join a multilateral written security guarantee to North Korea, if North Korea agreed to abandon its nuclear weapons programme. On 25 October, Pyongyang announced that it was prepared to consider 'written assurances of non-aggression' in place of a US–North Korea non-aggression treaty, as part of a 'simultaneous package solution' to the nuclear issue. On 4 November, KEDO formally 'suspended' construction of the LWR project for one year. This decision was long anticipated, but North Korea took no retaliation – beyond vowing that it would not allow any of the equipment or materials at the site to be removed.

Reaching agreement on a draft Six Party communiqué, however, proved difficult, as China sought to host another round of talks in mid-December. In early December, the US rejected a Chinese draft, which it thought was too generous to North Korea, and floated its own text, supported by South Korea and Japan, calling for 'coordinated steps', in lieu of North Korea's formula for 'simultaneous actions'. In response to the trilateral US–South Korean–Japanese draft, Pyongyang announced on 9 December that 'if the U.S. is concerned about the phraseology of simultaneous actions, we can accept an expression favored by the U.S. as long as there is no change in its content'. At the same time, however, Pyongyang said that it rejected US demands to disarm in exchange for 'written security assurances', arguing that it would only provide a 'word-for-word' commitment to disarm in exchange for security assurances. Pyongyang continued to insist that it would disarm in gradual stages, with benefits at every stage. On 9 December, North Korea also made public the details of its proposal, saying it would freeze all of its nuclear activities in exchange for the US lifting economic sanctions, removing

North Korea from the list of state sponsors of terrorism and resuming KEDO oil shipments. It was unclear, however, what activities would be covered by the freeze (in particular whether the enrichment programme was covered) and how it would be verified. In any event, Washington made clear that its near-term objective is dismantlement, not a freeze and that it was not prepared to 'reward' North Korea for restoring a freeze that it had broken in the first place. At the end of 2003, agreement could not be reached on a joint Six Party communiqué, although efforts continue to hold another round in 2004.

What future for the Six Party Talks?
The future of the Six Party Talks is uncertain. One possible scenario is that they will produce a new bargain to replace the Agreed Framework, requiring North Korea to disarm 'verifiably, completely, and irreversibly' in exchange for security assurances and political and economic benefits. All parties appear to accept the principle that any agreement will involve a staged or sequenced approach of simultaneous or coordinated steps, in which North Korea would disarm in a series of steps over a period of time, while the US and its allies take reciprocal steps at each stage along the way. North Korean disarmament steps could include accounting for plutonium produced prior to 1992, removing any spent fuel or separated plutonium from the country, dismantling the 5MW(e) reactor and destroying equipment and materials associated with the enrichment programme. Steps by the other parties could include a written security assurance, political normalisation of relations between Washington and Pyongyang, the lifting of economic sanctions, and providing various forms of economic and energy assistance.

In practice, however, working out the details of any sequenced series of steps appears very difficult to achieve over the next year, given the level of mistrust among the parties – especially between Washington and Pyongyang – and fundamental differences over issues of timing, sequencing and verification. As in the 1993–94 negotiations, North Korea has offered to accept a freeze of some sort, but not to give up its nuclear assets for the time being. In Pyongyang's view, it cannot afford to risk total disarmament while it still fears that Washington harbours ambitions to 'stifle' or 'strangulate' the regime. North Korean officials point to the fate of Saddam Hussein as an object lesson for why the regime needs to retain its nuclear deterrent. On the other side, the US appears willing to accept disarmament in stages, but not an agreement that allows North Korea to retain its nuclear capability for any period of time. In

'As in the 1993–94 negotiations, North Korea has offered to accept a freeze of some sort, but not to give up its nuclear assets for the time being'

Washington's view this would repeat the mistake of the Agreed Framework, leaving the US and its allies vulnerable to North Korean nuclear 'blackmail' in the future. As a result, negotiations for a sequenced solution are likely to run into difficulties as North Korea seeks to preserve as many of its nuclear assets as possible for as long as possible – in exchange for maximum benefits – while Washington tries to 'front-load' any agreement with near-term disarmament steps, while minimising benefits for the North until complete disarmament has been achieved.

Verification will also be a major stumbling block. Given its suspicion over North Korea's intentions, reinforced by Pyongyang's violation of the Agreed Framework and other agreements, Washington and other parties will insist on much more intrusive verification measures, which are certain to be strongly resisted by the secretive North Korean state. Reportedly, Washington, supported by Seoul and Tokyo, envisions teams of national experts, drawn from the Six Parties, with a mandate to conduct challenging inspections throughout North Korea. Even if the inspectors truly limit their search to suspect nuclear-related facilities and materials, they are bound to stumble across other secret military facilities, which, from Pyongyang's standpoint, would jeopardise its non-nuclear deterrent. Perhaps the only way to avoid extensive challenge inspections is for North Korea to take the initiative and volunteer information and allow access to secret facilities associated with its nuclear facilities.

Finally, procedural issues remain an obstacle. On paper, both the US and North Korea have accepted Six Party talks as the venue for negotiating an agreement, but underlying differences remain sharp. North Korea views the Six Party Talks as a political cover for intensive bilateral talks with the US, while Washington remains fundamentally resistant to serious bilateral negotiations with Pyongyang. Although the Six Party Talks have certain advantages for formalising multilateral security assurances and an assistance package, they are bound to be an unwieldy mechanism for give-and-take negotiations. Pyongyang hopes that Washington will agree to expert-level bilateral talks to hammer out the details of a sequenced approach, but Washington does not seem inclined at this stage to be drawn into such an approach.

If a diplomatic solution is not likely soon, a second possibility is that the talks will collapse. Washington could declare an end to diplomatic efforts and seek to impose international economic and political sanctions against North Korea, with the support of China, Russia,

South Korea and Japan. Always unpredictable, North Korea could carry out threats to conduct a missile test or a nuclear test or take some other action in an effort to increase pressure on the US to make concessions. For a variety of strategic and tactical reasons, however, all the participants in the Six Party Talks want to keep the negotiations alive for the time being and avoid the complications and risks of a breakdown. Despite its scepticism about North Korea's intentions, Washington cannot enlist support from key regional powers, or action by the UNSC to increase political and economic pressure on North Korea, unless it first demonstrates that Pyongyang has rejected a 'fair and reasonable' diplomatic solution. In any event, a confrontation with North Korea would strain US relations with its allies and China and further stretch US military forces, already burdened with a seemingly lengthy occupation of Iraq. Despite its threats, Pyongyang recognises that provocative actions intended to precipitate a crisis and wring concessions out of Washington would risk alienating China and facilitate US efforts to mobilise support for international sanctions and isolation that could cut off the foreign assistance vital for the regime's survival.

The third possibility is that the talks continue for the time being without a breakthough or a breakdown. In this scenario, neither Washington nor Pyongyang are prepared to make fundamental compromises in their respective positions, and each believes the other is constrained from forcing the issue. Washington believes that China can keep Pyongyang under control with the threat of ending assistance if North Korea misbehaves. Pyongyang calculates that Washington's focus on other issues, like Iraq, and pressure from other parties to keep the talks alive, will prevent Washington from walking out of the talks. As long as it continues to receive necessary outside assistance, Pyongyang may be content with protracted negotiations, while it awaits the outcome of the November 2004 US presidential elections. As long as North Korea does not force its hand, Washington may be content with a drawn out diplomatic process, so it can attend to more pressing and less divisive matters. While the other parties to the Six Party process would vastly prefer a solution, they are willing to settle for protracted talks that avoid a confrontation for the time being.

In such a scenario, compromise language could be found on a general set of principles to be issued at a future round of Six Party Talks, stressing joint commitment to achieve a nuclear-free Korean Peninsula and addressing security concerns and needs for economic development in the region. Further meetings

'For a variety of strategic and tactical reasons, however, all the participants in the Six Party Talks want to keep the negotiations alive for the time being and avoid the complications and risks of a breakdown'

could focus on how to implement these principles – a complex and difficult set of issues that is likely to require many rounds of discussions. Another difficult issue will be establishing the terms and conditions for restoring a freeze on North Korea's nuclear activities, so that North Korea cannot take advantage of protracted talks to enhance its nuclear capabilities. North Korea has proposed a freeze, but the critical details of what activities would be covered by the freeze and how it would be verified need to be determined. In theory, a freeze on the 5MW(e) reactor and on reprocessing could be monitored by the return of national or international inspectors to Yongbyon. Such inspectors could also account for any additional plutonium that North Korea might have separated in the meantime to verify that the raw separated plutonium had not been fabricated into nuclear weapons components. However, once the plutonium has been extracted from the spent fuel, it can be converted into nuclear weapons parts fairly easily and quickly. A preferable option would be to remove the plutonium to a third country for safe storage – an option that Pyongyang is not likely to find attractive. In any event, from Washington's perspective, a freeze that is limited to the plutonium programme – while allowing North Korea to proceed with its clandestine uranium enrichment efforts is no freeze at all since North Korea might be able to complete an enrichment facility and begin production of weapons-grade uranium even while plutonium production is frozen. If Pyongyang agreed to include uranium enrichment in the freeze, verification would be very challenging. Even if North Korea acknowledged the programme and volunteered unprecedented transparency, including access to plans, materials and sites, it will take considerable effort to determine whether everything has been declared and inspected, given the dearth of knowledge about the programme. A continuation of the Six Party Talks, as well as the intense bilateral diplomacy surrounding the formal meetings, is critical to begin to make incremental progress towards resolving these many complex and difficult issues, if a final agreement is to be reached.

Conclusion
For nearly 25 years, various diplomatic efforts have been underway to deal with the North Korean nuclear issue, whether to prevent North Korea from acquiring nuclear weapons or to disarm its assumed nuclear weapons capability. At different times, the instruments to achieve these objectives have included an international treaty (the NPT), a regional nuclear-free zone (the NSDD) and a bilateral agreement between the US and North Korea (the Agreed Framework). In the end, none of these diplomatic efforts and agreements have been fully successful, although, in different degrees, they have helped to delay or constrain – sometimes very significantly – North Korea's nuclear weapons efforts. The current diplomatic

effort involves a fourth variation – a Six Party multilateral agreement or perhaps a package of bilateral agreements stitched together in an overall multilateral framework. As with previous efforts, the future success or failure of the Six Party Talks is uncertain.

In crafting their approaches to the nuclear issue, the US and other powers have struggled to come to grips with Pyongyang's ultimate intentions. For years, North Korea watchers have debated whether Pyongyang views nuclear weapons as indispensable to the regime's survival and therefore non-negotiable, or whether it sees its nuclear assets as a bargaining chip to be traded away for political and economic benefits necessary to sustaining the regime. The historical record suggests that the answer is both, and the emphasis that Pyongyang places on one or the other varies with domestic conditions and external circumstances.

On the one hand, the time and energy that North Korea has invested in developing its nuclear weapons capability, allied to its willingness to repeatedly violate nuclear agreements, strongly suggests that North Korean leaders deeply believe that some kind of nuclear hedge – or at least the appearance of a credible nuclear hedge – is essential to regime survival. Pyongyang sees itself as a besieged and beleaguered state, surrounded by more powerful enemies, untrustworthy allies, and a spectacularly successful southern competitor. In this view, if North Korea is ever going to reform itself and survive in the long run, it must find respite from external pressures and perceived threats. For such a state, nuclear weapons are the ultimate defence. As long as outside powers believe that it has a nuclear deterrent they are – in Pyongyang's view – more likely to leave North Korea alone and less likely to pursue hostile policies that could provoke a confrontation in which such weapons are used.

On the other hand, North Korea has demonstrated that it does respond to international inducements and pressures to limit its nuclear programme. In the past, Soviet diplomacy, backed by promises of nuclear power assistance, persuaded North Korea to join the NPT, and US diplomatic efforts convinced North Korea to implement IAEA inspections. Later, Washington and Pyongyang negotiated a complex bilateral agreement that froze North Korea's plutonium production facilities and established a process for the eventual elimination of these facilities. With a combination of carrots and sticks, Washington convinced Pyongyang to open up a secret underground facility and to accept a moratorium on long-range missile tests. Even when North Korea has created a crisis – by threatening to withdraw from the NPT, threatening to reprocess nuclear material, or threatening to test missiles – it typically has sought to leave the door open for a diplomatic exit. In short, diplomatic efforts in the past have constrained, but not eliminated North Korea's nuclear capabilities.

Nuclear diplomacy timeline (1980–2003)

1980

The US detects construction of a new reactor at the Yongbyon Nuclear Research Centre.

12 December 1985

North Korea accedes to the Treaty on the Non-Proliferation of Nuclear Weapons (the nuclear Non-Proliferation Treaty or NPT).

13 March 1987

North Korea accedes to the Convention on the Prohibition of the Development, Production and Stockpiling of Bacteriological (Biological) and Toxin Weapons and on Their Destruction (the Biological Weapons Convention, or BWC).

27 September 1991

President George H.W. Bush announces that US tactical nuclear weapons would be removed from overseas locations, including South Korea.

31 December 1991

North and South Korea conclude the North–South Denuclearization Declaration (NSDD).

30 January 1992

North Korea signs a full-scope safeguards agreement with the International Atomic Energy Agency (IAEA). The agreement enters into force on 10 April 1992.

26 May 1992

IAEA inspections begin. Sample analysis suggests that more plutonium has been produced than has been declared; the US detects apparent efforts to conceal underground waste sites.

11 February 1993

The IAEA formally requests special inspection of suspect waste sites.

25 February 1993

The IAEA Board of Governors gives North Korea a one-month deadline to accept special inspections.

12 March 1993

North Korea announces its intention to withdraw from the NPT.

1 April 1993

The IAEA Board of Governors reports North Korea's violation of the NPT to the UN Security Council (UNSC).

11 May 1993

The UNSC adopts Resolution 825. This leads to negotiations between the US and North Korea.

29 May 1993

North Korea tests a *No-dong* missile.

21 October 1994

The US and North Korea sign the Agreed Framework.

9 March 1995

The US, South Korea and Japan establish the Korean Peninsula Energy Development Organization (KEDO).

15 December 1995

KEDO and North Korea conclude the Light-Water Reactor (LWR) supply agreement.

19 December 1997

Kim Dae Jung wins the South Korean elections, and institutes the 'sunshine policy'.

17 August 1998

Press reports note suspect nuclear facility at Kumchang-ri.

31 August 1998

North Korea launches a *Taepo-dong* missile.

17 March 1999

North Korea agrees to a US visit to Kumchang-ri in May. The visit shows that the underground structure is not intended for nuclear-related uses.

Disarmament Diplomacy with North Korea

Nuclear diplomacy timeline (1980–2003) continued

7–12 September 1999
North Korea agrees to a moratorium on further long-range missile tests.

6 June 2001
The Bush administration of President George W. Bush publishes its 'broad agenda' policy towards North Korea.

16 October 2002
The US announces North Korea's clandestine enrichment programme.

14 November 2002
KEDO suspends further oil shipments to North Korea.

12 December 2002
North Korea announces the restart of its 5MW(e) reactor.

19 December 2002
Roh Moo Hyun wins the South Korean elections and promises to continue his predecessor's 'sunshine policy'.

27 December 2002
North Korea expels IAEA inspectors and announces that it will soon complete preparations to resume operations at the reprocessing facility.

6 January 2003
The IAEA Board of Governors calls on North Korea to allow the return of inspectors.

10 January 2003
North Korea announces its withdrawal from the NPT.

12 February 2003
The IAEA Board of Governors reports North Korea's NPT violation to the UNSC.

18 April 2003
North Korea announces that it is in the final phase of successfully reprocessing more than 8,000 spent fuel rods.

24–25 April 2003
Three Party Talks, between the US, China and North Korea are held in Beijing.

27–29 August 2003
The first round of Six Party Talks (between the US, Russia, China, Japan, and North and South Korea) are held in Beijing.

2 October 2003
North Korea publicly announces that it has successfully finished reprocessing the 8,000 spent fuel rods and is using the resulting plutonium to increase its 'nuclear deterrent force'.

19 October 2003
President George W. Bush indicates that the US is prepared to give written security assurances to North Korea in exchange for complete, verifiable, and irreversible disarmament.

4 November 2003
KEDO formally suspends construction of the LWR project for one year.

9 December 2003
North Korea announces a proposal for a nuclear freeze.

North Korea's Nuclear Weapons Programme

Overview

North Korea's nuclear efforts can be divided into four distinct phases. During the first phase (1959–80), the country's nuclear programme was primarily focused on basic training and research. North Korea relied on assistance from the Soviet Union, which trained North Korean scientists and engineers and helped to construct basic research facilities – including a small research reactor and a radioisotope production laboratory. These facilities were placed under International Atomic Energy Agency (IAEA) safeguards in 1977.

The second phase (1980–94) covers the growth and eventual suspension of North Korea's indigenous plutonium production programme. Around 1980, Pyongyang embarked on a major campaign to build a series of industrial-scale nuclear facilities that could produce substantial amounts of nuclear energy and weapons-grade plutonium. Although North Korea acceded to the 1968 nuclear Non-Proliferation Treaty (NPT) in December 1985, it denied the existence of these facilities until 1992, when it finally concluded a safeguards agreement with the IAEA permitting full inspections. Subsequently, North Korea's refusal to cooperate with the IAEA and to account for possible plutonium production prior to 1992 led to the 1993–94 nuclear crisis and, eventually, to the 1994 US–North Korea Agreed Framework. Although this accord froze North Korea's plutonium production facilities and placed them under IAEA monitoring, the US estimated that Pyongyang could have recovered enough plutonium for one or two nuclear weapons before this agreement came into force. The actual amount of plutonium produced by North Korea before 1992 is unknown.

The third phase (1994–2002) covers the period of the nuclear freeze. During these eight years, North Korea's indigenous plutonium production facilities remained in a state of suspended animation, and its known plutonium stocks – some 25–30 kilograms (kg) in spent-fuel rods – were subject to IAEA monitoring. With the plutonium route blocked by the Agreed Framework, which also required North Korea to eventually declare the plutonium produced prior to 1992, Pyongyang instigated a secret programme in the late 1990s to develop the means to produce weapons-grade enriched uranium utilising gas centrifuge technology.

The final phase examines the period from the end of the freeze in late 2002, following the exposure of Pyongyang's secret uranium enrichment programme, to the present day. North Korea has taken some initial steps to revive its plutonium production facilities after nearly a decade of dormancy, and claims it has extracted all the plutonium it has on hand in spent fuel rods (though this cannot be confirmed by independent means). Presumably, North Korea is also continuing with its efforts to complete development of a significant uranium enrichment capability. However, very little is known about this project.

Phase 1 – origins of North Korea's nuclear programme

North Korea's nuclear programme was born with assistance from the Soviet Union. The two countries signed a nuclear cooperation agreement in 1959 and, over the next 30 years, Moscow provided Pyongyang with training and technology useful in the development of basic nuclear technology. The type of aid granted to North Korea was typical of that on offer during the Cold War, when both the Soviet Union and the US supplied some of their allies and client states with basic nuclear technology and training. The 1959 agreement enabled a variety of technical and scientific exchanges and projects, including construction of the Yongbyon Nuclear Research Centre, training of North Korean scientific and technical personnel, and geological surveys for nuclear applications. Soviet assistance was not specifically intended to assist the development of nuclear weapons, but it allowed Pyongyang to master the basic technologies needed to produce and separate plutonium, which North Korea later employed in its nuclear-weapons programme.

In the early 1960s, with Soviet assistance, North Korea began construction of the Yongbyon Nuclear Research Centre, which became the centrepiece of its nuclear programme. Initially, the principal facilities housed at Yongbyon comprised a small research reactor, the IRT-2000, designed to conduct basic nuclear research and to produce small quantities of medical and industrial isotopes, and an adjacent radiochemical laboratory for extracting isotopes from 'targets' irradiated in the IRT-2000. The IRT-2000 is a 'pool-type' research reactor fuelled by a mixture of fuel elements of 10%, 36% and 80% enriched uranium, moderated and cooled by 'light' (i.e. ordinary) water.[1] Construction of the IRT-2000 began in 1963. It became operational in 1965 at a power rating of 2MW(th), which was upgraded to 4MW(th) in 1974, and to 8MW(th) in 1987.

The radiochemical laboratory, which became operational in 1977, was originally fitted with 20 shielded hot cells and glove boxes, used to process radioactive isotopes for medical and industrial

'North Korea's nuclear programme was born with assistance from the Soviet Union' ... 'Soviet assistance was not specifically intended to assist the development of nuclear weapons, but it allowed Pyongyang to master the basic technologies needed to produce and separate plutonium'

North Korea's Nuclear Weapons Programme

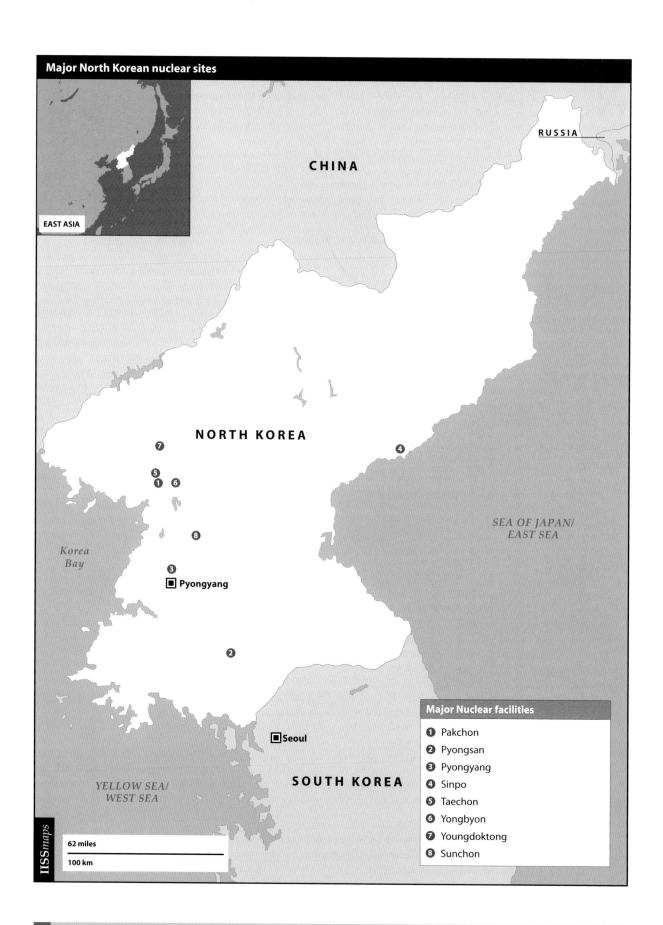

Major North Korean nuclear sites

EAST ASIA

RUSSIA

CHINA

NORTH KOREA

❼

❺
❶ ❻

❽

Korea
Bay

❸
▣ Pyongyang

SEA OF JAPAN/
EAST SEA

▣ Seoul

SOUTH KOREA

YELLOW SEA/
WEST SEA

62 miles

100 km

IISS*maps*

Major Nuclear facilities

❶ Pakchon
❷ Pyongsan
❸ Pyongyang
❹ Sinpo
❺ Taechon
❻ Yongbyon
❼ Youngdoktong
❽ Sunchon

Major North Korean nuclear sites	
1. Pakchon	Location of a uranium mine and milling facility (known as the April Industrial Enterprise), declared to the IAEA in 1992. The uranium milling facility reportedly processes ore from mines in the Sunchon area. Current status is unknown.
2. Pyongsan	Location of uranium mining and a uranium milling facility, declared to the IAEA in 1992. The milling facility in Pyongsan reportedly processes ore from the Pyongsan and Kumchon uranium mines. Current status is unknown.
3. Pyongyang	Laboratory-scale hot cells, provided by the Soviet Union in the 1960s.
4. Sinpo	Location of two 1,000 MW(e) light water reactors being constructed by the Korean Energy Development Organization (KEDO) under the terms of the Agreed Framework, which set a 'target date' of 2003 for completion of the project. Various events have delayed the project. Construction began in mid-1997. The major non-nuclear element for the first reactor, defined in the Agreed Framework as a 'significant portion' of the LWR project, was scheduled for completion by mid-2005. KEDO 'suspended' the project for one year in December 2003.
5. Taechon	Location of an incomplete 200MW(e) graphite-moderated nuclear power reactor. Construction began in 1989 and was frozen in 1994 (under the 1994 Agreed Framework). Current status is unknown, but there are no reports of major construction resuming after North Korea renounced the nuclear freeze in December 2002.
6. Yongbyon	Location of a Nuclear Research Centre, comprising a 5MW(e) graphite moderated research nuclear power reactor, an unfinished 50MW(e) graphite moderated prototype power reactor, reprocessing facility, uranium conversion plant, fuel fabrication plant, and spent fuel and waste storage facilities. Also location of a Soviet-supplied IRT research reactor and radioisotope laboratory. Operation of the 5MW(e) reactor, the uranium conversion plant, the fuel fabrication facility and the reprocessing plant were frozen in 1994, along with construction of the 50MW(e) reactor. Since 2002, North Korea has restarted the 5MW(e) reactor and reportedly reprocessed some or all of the 8,000 spent fuel rods at the site. No resumption of work on the 50MW(e) reactor has been reported.
7. Youngdoktong	Reported location of site (active in the 1990s) for nuclear weapons-related high-explosive testing.
8. Sunchon	Location of an important uranium mine. Other mines reportedly located in Kumchon, Pyongsan, and Hwangsan.

Note: There are assumed to be several additional nuclear facilities associated with North Korea's enrichment programme, including a possible production-scale centrifuge plant. The location of these facilities is unknown.

Sources: Carnegie Endowment for International Peace, www.ceip.org; Federation of American Scientists, www.fas.org; Nuclear Threat Initiative, www.nti.org; and David Albright and Kevin O'Neill, *Solving the North Korea Nuclear Puzzle*, (Washington, DC: The Institute for Science and International Security, 2000).

Central area of Yongbyon nuclear complex

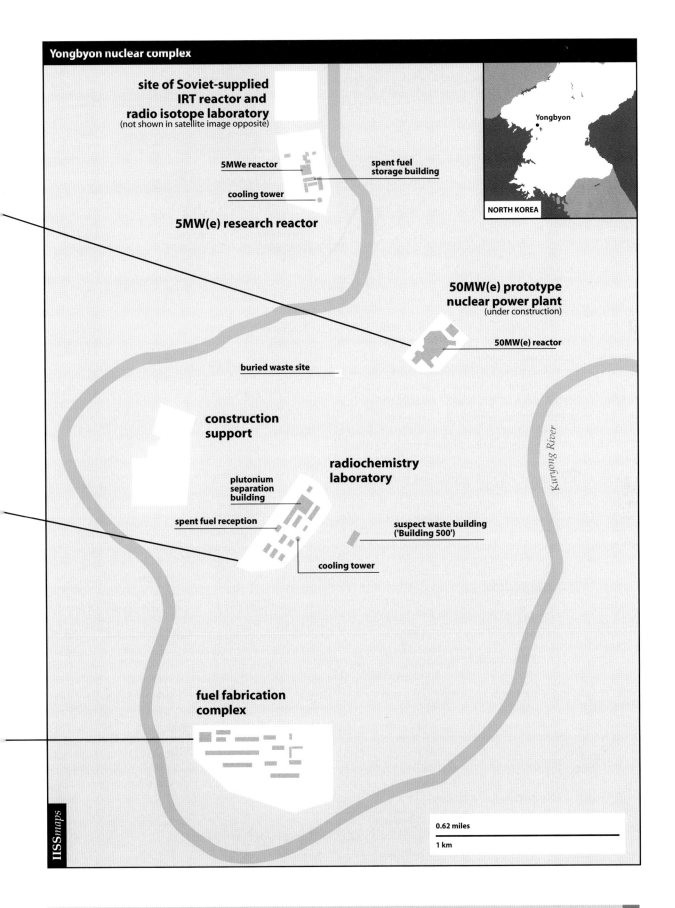

Yongbyon nuclear complex

site of Soviet-supplied
**IRT reactor and
radio isotope laboratory**
(not shown in satellite image opposite)

5MWe reactor

cooling tower

spent fuel
storage building

5MW(e) research reactor

**50MW(e) prototype
nuclear power plant**
(under construction)

50MW(e) reactor

buried waste site

**construction
support**

**radiochemistry
laboratory**

plutonium
separation
building

spent fuel reception

suspect waste building
('Building 500')

cooling tower

**fuel fabrication
complex**

Yongbyon

NORTH KOREA

Kuryong River

0.62 miles

1 km

IISS*maps*

purposes from irradiated targets. Typically, target materials were irradiated in the research reactor and then transferred to the hot cells and glove boxes where the desired isotope was chemically separated from waste products. Soviet experts also assisted in the construction of an underground facility at the site for storing radioactive waste from isotope production.

Although the IRT-2000 reactor and the radiochemical laboratory were intended for basic nuclear research and isotope production, the materials and equipment also provided North Korea with the means to experiment with the production and extraction of small amounts of plutonium, which Pyongyang acknowledged to the IAEA in 1993. These Soviet-supplied facilities were placed under IAEA inspections in 1977. However, due to IAEA procedures for monitoring facilities of this type, they were not subject to close scrutiny. The total amount of plutonium produced is, therefore, uncertain.

Although Soviet training of North Koreans is reported to have begun a few years before 1959, under the 1959 agreement and subsequent arrangements, the Soviet Union trained more than 300 (an exact figure regarding the number of experts who received training and in what fields is not publically available) North Korean engineers and physicists at Soviet institutes, including the Joint Institute for Nuclear Research at Dubna, the Moscow Engineering Physics Institute, and the Bauman Higher Technical School. (North Korean nuclear experts also received training in Canada, Japan and the former East Germany.)

Meanwhile, geological surveys conducted by the Soviet Union determined that North Korea possessed large deposits of uranium ore and graphite – Pyongyang subsequently developed these to form the building blocks of its plutonium production programme.

Phase 2 – North Korea's plutonium production programme

Around 1980, North Korea launched a concerted national programme to build a series of industrial-scale facilities capable of producing significant amounts of plutonium for a nuclear-weapons programme, as well as for the country's nuclear-power industry. The core of this programme were three gas-cooled, graphite-moderated, natural-uranium-fuelled reactors:

- a small 5MW(e) (25MW(th)) research reactor at Yongbyon;
- a larger 50MW(e) (200MW(th)) prototype power reactor at Yongbyon; and

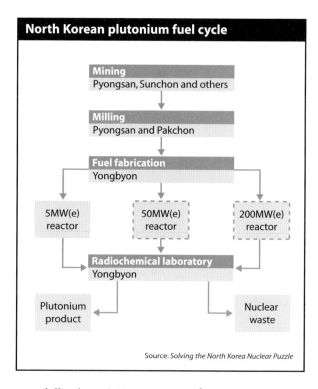

North Korean plutonium fuel cycle

Mining
Pyongsan, Sunchon and others

Milling
Pyongsan and Pakchon

Fuel fabrication
Yongbyon

5MW(e) reactor | 50MW(e) reactor | 200MW(e) reactor

Radiochemical laboratory
Yongbyon

Plutonium product

Nuclear waste

Source: *Solving the North Korea Nuclear Puzzle*

- a fullscale 200MW(e) (800MW(th)) power reactor at Taechon.

The power levels of nuclear reactors can be measured in terms of megawatts of heat produced (MW(th)) or megawatts of electricity generated (MW(e)). Plutonium production capacity is a function of thermal power, but North Korea always designated its reactors according to their electrical output, presumably to emphasise their civil intent. As a result, North Korea's reactors are usually identified by their nominal electrical output capacity, a method followed in this document.

Around this trio of reactors, North Korea also constructed facilities for the full plutonium fuel cycle. At the 'front end' of the fuel cycle were uranium mines, factories to process and refine uranium ore, as well as plants to purify natural uranium, to convert it to metal, and to fabricate fuel. At the 'back end' of the fuel cycle was an industrial-scale reprocessing plant at the Yongbyon Nuclear Research Centre designed to extract plutonium from spent reactor fuel along with facilities to treat and store nuclear waste.

The technology that North Korea chose was attractive for several reasons. Based on 1950s technology originally developed by France and the UK (to produce plutonium for their nuclear-weapons programmes),

'Around 1980, North Korea launched a concerted national programme to build a series of industrial-scale facilities capable of producing significant amounts of plutonium for a nuclear-weapons programme, as well as for the country's nuclear-power industry'

the basic reactor designs were available in the public domain and relatively straightforward to build and operate. Since the raw materials for these reactors – large quantities of natural uranium and graphite – could be found locally, Pyongyang was able to pursue an indigenous nuclear programme with minimal dependence on foreign assistance.

In addition, these reactor designs are well suited to producing weapons-grade plutonium.[2] In contrast to light-water reactors (LWRs), which use low enriched uranium, graphite-moderated reactors fuelled by natural uranium are relatively more efficient in producing plutonium. In addition, the fuel is designed for moderately low levels of irradiation or 'burn up', minimising the accumulation of certain plutonium isotopes that are undesirable for weapons purposes. Moreover, the type of fuel used in the graphite-moderated reactors developed by North Korea is generally unsuitable for long-term storage because the metal cladding surrounding the uranium fuel rods tends to corrode rapidly in water and presents a fire hazard if exposed to air. Consequently, the irradiated fuel is usually reprocessed – that is, the plutonium and uranium are chemically separated from radioactive waste products, which can be safely stored in specialised tanks and containers.

Finally, just as graphite-moderated reactor technology was initially developed for military production but later adapted for energy production in western countries, this choice of reactor provided North Korea with a 'dual-use' capability – technology that could be utilised in nuclear-weapons production as well as nuclear-power production.

The main facilities that comprise North Korea's plutonium production complex are detailed below.[3]

Uranium mining and milling

On the basis of geological surveys conducted by the Soviet Union, North Korea began large-scale uranium mining operations at various locations near Sunchon and Pyongsan in the late 1970s or early 1980s. The raw uranium-bearing ore was shipped to uranium milling factories at Pakchon and Pyongsan, where it was crushed and chemically processed to produce U_3O_8 or 'yellow cake', which was then transported to the Yongbyon nuclear centre for further processing and fabrication into nuclear reactor fuel. Typically, one tonne of North Korean uranium ore contains about one kilogram of uranium, meaning that some 50,000 tonnes of ore had to be mined and processed in order to obtain the 50 tonnes of natural uranium needed for the initial fuel load for the 5MW(e) reactor. It has been estimated that, at its peak in the early 1990s, North Korea was able to produce about 300 tonnes of yellow cake annually, equal to approximately 30,000 tonnes of uranium ore. Actual production of yellow cake in the

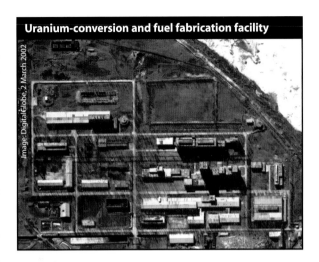
Uranium-conversion and fuel fabrication facility
Image: DigitalGlobe, 2 March 2003

decade before the nuclear freeze is unknown. North Korea's current mining and milling capacity is also unknown, but it appears unlikely that yellow cake production is a significant constraint on its immediate nuclear requirements.

Uranium conversion and fuel fabrication

Between 1980 and 1985, North Korea built a substantial factory at Yongbyon to refine yellow cake and to produce uranium metal fuel elements for its graphite-moderated reactors. The yellow cake was chemically refined in a series of buildings using a process known as 'conversion' into more purified forms of uranium, from U_3O_8 to UO_3 (uranium trioxide) to UO_2 (uranium dioxide). At a high temperature, the UO_2 was mixed with highly caustic gaseous hydrofluoric acid to produce a more purified uranium compound – uranium tetrafluoride (UF_4) – which was converted into uranium metal ingots in vacuum furnaces. In the final stage of fuel fabrication, the uranium metal ingots were melted and alloyed with small amounts of aluminum, with the resulting alloy extruded and machined into fuel rods of around three centimetres (cm) in diameter and 50cm in length. Finally, each uranium fuel rod was inserted into a magnesium-zirconium alloy cladding tube (known as 'magnox') to produce the final fuel assembly. Approximately 8,000 such fuel assemblies, containing about 50 tonnes of natural uranium, were necessary for the 5MW(e) reactor core. If completed, the larger 50MW(e) reactor would require about 400 tonnes of uranium fuel, equal to some 64,000 fuel assemblies, while the 200MW(e) reactor would require about 1,400 tonnes of uranium fuel, equal to about 224,000 fuel assemblies.

The uranium-conversion and fuel-fabrication facility at Yongbyon was designed to produce fuel for the entire line of graphite-moderated reactors under construction in North Korea during the 1980s. However, the level of production before the nuclear freeze is unknown. According to North Korean officials, in 1992, the plant was producing roughly 100 tonnes of uranium fuel per annum, equal to approximately 16,000 fuel assemblies, although its nominal annual capacity was larger, perhaps 200–300 tonnes of uranium fuel (32,000–48,000 fuel assemblies). At a minimum, North Korea in known to have produced enough fuel prior to the freeze for the initial core load for the 5MW(e) reactor and at least one fresh core load. It is also known to have produced slightly more than one-half of the fuel required for the 50MW(e) reactor under construction. North Korea could have produced a significant amount of additional fuel before 1992 that it failed to declare to the IAEA.

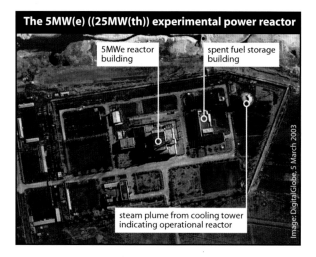

The 5MW(e) ((25MW(th)) experimental power reactor

5MWe reactor building

spent fuel storage building

steam plume from cooling tower indicating operational reactor

Image: DigitalGlobe, 5 March 2003

5MW(e) [(25MW(th)] experimental power reactor[4]
During the 1980s, the most important facility in North Korea's plutonium production programme was a small research reactor located at the Yongbyon Nuclear Research Centre, which was designated an 'experimental power reactor'. Based on the same design concept as the UK's 50MW(e) Calder Hall plutonium production reactor, which became operational in 1956, the North Korean reactor is fuelled with magnox-clad, natural-uranium fuel elements cooled with carbon-dioxide gas (CO_2) and moderated and reflected with high purity graphite. In design, the reactor core consists of some 300 tonnes of graphite blocks into which 812 fuel channels have been drilled vertically. Each fuel channel is designed to hold ten fuel assemblies stacked vertically on top of one another, giving a total of about 8,000 fuel rods (50 tonnes of natural uranium) in a full core load. In addition to the fuel channels, some 40

control channels have been drilled into the graphite blocks to allow for the insertion of barium carbon control rods to maintain control over the nuclear reaction. The graphite blocks holding the fuel and the control-rod channels are surrounded by an additional 300 tonnes of graphite reflector (which serves to reflect neutrons back into the core) and the entire core is encased in a steel pressure vessel to contain the cooling gas. Cooling takes place using pressurised CO_2, which is blown through the core by electric motors. A large loading and unloading machine refuels the reactor from the top of the core.

Reactor construction began in 1980, and the reactor went critical in August 1985. It operated intermittently from 1986 until 1994 when it was shut down under the Agreed Framework. According to North Korea, the reactor was designed to have a power rating of 25MW(th), which North Korea expressed in terms of its nominal 5MW(e) electrical output, although the reactor was not actually used to produce electricity. In theory, operating at full power for 300 days per year, this reactor could produce approximately 7.5kg of weapons-grade plutonium annually in the discharged spent fuel.[5] Of course, actual annual plutonium production would depend on the fuel's irradiation level, which is a function of the reactor's power level and the number of days per year that it was operational, usually expressed as megawatt thermal days per tonne (MW(th-d/t)).

The operational history of the reactor between 1986 and 1994 (and hence how much plutonium was produced) is shrouded in mystery. In 1992, when IAEA inspectors were first allowed access to the reactor, Pyongyang claimed that the reactor had experienced serious start-up and control difficulties, which prevented full-power operations and resulted in frequent shut downs during the first several years of operations. According to the North Koreans, reactor operation caused distortions in the neutron flux (the total number and speed of neutrons within a specific volume), which caused fuel in some parts of the core to overheat and to fail. In April–May 1989, the reactor was reportedly shut in order to remove a few hundred damaged fuel rods, after which North Korea claimed that it was able to operate more regularly at 20MW(th) – close to full power. At the time of the initial IAEA inspections in April 1992, North Korea claimed that about 17kg of plutonium had been produced in the reactor fuel. By the time the reactor was completely de-fuelled in June 1994, the spent fuel probably contained about 25–30kg of plutonium.

'By the time the reactor was completely de-fuelled in June 1994, the spent fuel probably contained about 25–30kg of plutonium'

The accuracy of North Korea's operating record prior to 1992 has never been verified, and independent evidence is mixed. Through satellite reconnaissance, the US detected the construction of the 5MW(e) reactor at an early stage, and was able to confirm initial operation of the reactor in 1986 by noting the emission of steam plumes from its cooling tower, which indicated the reactor was venting excess heat. However, satellite monitoring of the facility was not frequent enough during the 1980s to compile a complete record of operations and cloud cover sometimes prevented plumes from being detected. In any event, 'plumology' is an inexact science. If operated at low power levels or under certain climatic conditions, the reactor might not produce a clearly visible steam plume. Nonetheless, US observation of the reactor over the years did reveal some discrepancies in North Korea's declared operating

The 200MW(e) (800MW(th)) power reactor

Image: DigitalGlobe, 3 September 2002

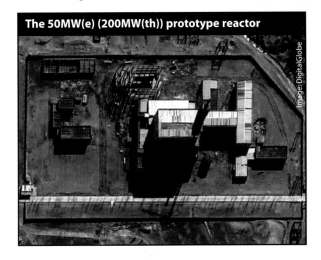

The 50MW(e) (200MW(th)) prototype reactor

Image: DigitalGlobe

history, suggesting that the reactor may have functioned more frequently than Pyongyang had claimed.

The IAEA was also never able to complete its investigations into the reactor's operating history. Some of the evidence collected by the Agency supported North Korea's assertions, while some of it did not. Furthermore, Pyongyang obstructed the IAEA's efforts to take further measurements that might have provided answers to these questions. In fact, North Korea deliberately destroyed evidence which would have been helpful in reconstructing and verifying the reactor's operating history. In June 1994, when North Korea completely de-fuelled the reactor, it refused to allow the IAEA to record the exact location of each of the 8,000 fuel rods in the core. Armed with this information, and subsequent measurements of the radioactivity level of each rod, the IAEA had hoped to 'map' the reactor's operational history, especially to determine how long the fuel was in the reactor. To foil this effort, North Korea not only prevented the IAEA measurements, but also mixed the rods together in

different storage baskets in the cooling pond so that the evidence was destroyed.

The 50MW(e) (200MW(th)) prototype power reactor

In 1984, North Korea began construction of a larger reactor at Yongbyon, using the same basic materials and technology as utilised in the 5MW(e) reactor – magnox-clad natural uranium fuel, graphite moderation, and CO_2 gas cooling – although the core design concept in this case resembles the French-designed G2 reactor commissioned at Marcoule in 1956 with fuel rods placed horizontally rather than vertically, as in the Calder Hall-type design. Nominally rated at 50MW(e) and with a core load of 400 tonnes of natural uranium fuel this reactor is, theoretically, capable of producing about 55kg of weapons-grade plutonium per year, if operated at full power for 300 days and assuming a discharge of 100 tonnes of the most heavily irradiated spent fuel. According to North Korean officials, the reactor was within a year of initial service at the time of the nuclear freeze, although this claim was never verified. Because it was never completed, it is unknown whether the reactor is capable of full-power operations, and there are no clear signs that contruction on the reactor has resumed since the nuclear freeze ended in 2002.

The 200MW(e) (800MW(th)) power reactor

In the late 1980s, North Korea began construction at Taechon of a fullscale version of the 50MW(e) reactor, based on the same technology – magnox-clad natural-uranium fuel, graphite moderation, and CO_2 gas cooling – and the same core design as the French G2. Nominally rated at 200MW(e), and with a core load of 1,400 tonnes of uranium fuel this reactor would, theoretically, be capable of producing up to 220kg of weapons-grade plutonium annually, if operated at full power for 300 days per year. This reactor was in the early stages of construction when the nuclear freeze came into effect in 1994.

Radiochemical laboratory/reprocessing plant

In 1984, North Korea began construction of an industrial-scale reprocessing plant to separate plutonium from spent nuclear fuel at the Yongbyon Nuclear Research Centre. During construction, the exact purpose of the facility was debated within the US intelligence community. Some analysts believed that it was most likely a reprocessing facility, while others argued that it could be engaged in non-nuclear industrial activities. It was not until the IAEA conducted inspections in 1992 that it was confirmed as a reprocessing plant, which Pyongyang euphemistically called a 'radiochemical laboratory'. The operation of the plant is based on the Purex (plutonium uranium extraction) process, a standard procedure widely used in the nuclear industry, in which uranium and plutonium are separated from nuclear waste and then from each other through a series of chemical processes. Lorries transport spent fuel from the 5MW(e) reactor in buckets stored inside heavy casks to the front end of the reprocessing facility, where the nuclear-fuel assemblies are mechanically dis-assembled, and the uranium fuel is dissolved in nitric acid. The liquid mixture is treated with various chemicals and passed through a series of stainless-steel mixer settler tanks in which the plutonium and uranium is selectively precipitated from the spent fuel's highly radioactive fission products. The aqueous plutonium-uranium mixture is then passed through another set of mixer settler tanks to separate out the plutonium. Because of high radiation, all of the operations are carried out remotely behind heavy shielding. Finally, at the back end of the plant, the concentrated plutonium is further purified and collected as plutonium oxide ($PuO4$) in a series of glove boxes. The oxide powder can then be converted into plutonium metal ingots, which could be melted and cast into components for nuclear weapons. Adjacent to the main building are a series of tanks and vaults intended to concentrate and store the large volumes of liquid and solid radioactive waste produced during reprocessing.

In 1992, IAEA inspectors discovered that one reprocessing line had been completed at the plant and that a second was under construction. According to North Korean officials, at this time, the reprocessing facility was designed to process spent fuel containing 0.7kg of plutonium per tonne of spent fuel, and each line was designed to process one tonne of spent fuel over the course of about three days of continuous operations. During reprocessing, plants typically operate around-the-clock, although it is possible (if less efficient) to process individual batches of spent fuel one at a time. Theoretically, the facility's one completed line is capable of processing the 5MW(e) reactor's entire 50 tonne core load in a single campaign, lasting approximately 150 days. If both lines were operating continuously for 300 days per year, the plant would have a total nominal capacity to process annually some 200 tonnes of magnox spent fuel, more than sufficient to handle the spent fuel that would typically be discharged each year by the 5MW(e) and the 50MW(e) reactors. In 1994, when IAEA inspectors returned to monitor the nuclear freeze, they found that North Korea had made considerable progress in installing equipment for the second reprocessing line, which was scheduled for completion in 1996.

The 1992 plutonium mystery

North Korea's accession to the NPT in December 1985 necessitated that it place all its nuclear facilities and materials under international inspection and that it pursue nuclear technology solely for peaceful purposes. Although North Korea had 18 months under the treaty to negotiate a comprehensive safeguard agreement with the IAEA, it did not sign this agreement until January 1992. In all, six official inspection missions took place in North Korea in 1992, before Pyongyang denied inspectors access to suspect nuclear waste storage facilities and threatened to withdraw from the NPT.

During the first inspection in 1992, North Korea told the IAEA that it had test run the reprocessing plant between March–May 1990, during which 86 damaged fuel rods (which had been removed from the 5MW(e)

Radiochemical laboratory/reprocessing plant

Image: DigitalGlobe

high level waste

chemical storage building

analytical labs

plutonium separation building

spent fuel reception

'In 1984, North Korea began construction of an industrial-scale reprocessing plant to separate plutonium from spent nuclear fuel at the Yongbyon Nuclear Research Centre'

reactor in 1989), as well as 172 fresh fuel rods, were reprocessed in a single campaign of three batches. According to North Korea, this resulted in the recovery of 62 grams of plutonium oxide from the nearly 90 grams of plutonium contained in the spent fuel, a claimed high loss rate of approximately 33%.[6] To verify the accuracy of North Korea's declaration, the IAEA took samples from the extracted plutonium, as well as from the waste tanks and work areas at the reprocessing plant, including 'swipe samples' from the glove boxes in which the plutonium was processed. Such samples contained individual dust particles that could be analysed to deduce the fractional content and ratio of different isotopes produced in the irradiated fuel.

Analysis of these samples highlighted several discrepancies in North Korea's initial declaration to the IAEA. The Plutonium-240 (Pu-240) content of the declared plutonium was uniform, but the dust particles contained a range of Pu-240 content that tended to cluster in three distinct groups, suggesting that additional batches of fuel (containing slightly different levels of Pu-240) had been reprocessed. Contrary to North Korea's declaration that it had only conducted reprocessing in 1990, the ratio of Americium-241 to Pu-241 found on these dust particles at the site suggested that at least three reprocessing campaigns had occurred, in 1989, 1990 and 1991. Finally, the IAEA discovered that the percentage of Pu-240 in the declared plutonium was different to that in the waste tanks, suggesting that there was a quantity of 'missing' nuclear waste. To explain these discrepancies, North Korean officials said that a a few hundred milligrams of plutonium had been separated in 1975 from targets irradiated in the IRT-2000 reactor, and that the waste from these experiments had been mixed with the waste from the 1990 test run, contaminating the IAEA's findings. This explanation did not seem plausible to IAEA analysts, because the additional waste declared by North Korea seemed too old and too small to explain the discrepancies.

As the IAEA began to discover these discrepancies, the US reviewed the record of satellite imagery of the 5MW(e) reactor. Based on the absence of steam plumes from the cooling tower, the reactor appeared to have been shut down for about two months in early 1989, as North Korea claimed. During this time, North Korea contended that it had replaced some 300 fuel rods damaged by overheating due to flaws in the reactor's design. Of these damaged fuel rods, North Korea said that 86 were subsequently reprocessed, and that the others were stored in a dry storage pit at the reactor site. US analysts, though, estimated that, in the worst case, North Korea could have unloaded a much larger portion or even the entire core load during the two-month shut down. This conclusion was based on a calculation of how quickly one or two fuel machines could unload spent fuel and load fresh fuel, assuming

the North Koreans worked around-the-clock.[7] Based on Pyongyang's account of the reactor's operating history prior to the 1989 fuel discharge, the entire core would have contained about 9.5kg of plutonium, which, assuming a likely range of potential 10–30% reprocessing losses, could have yielded as much as 6.5–8.5kg of plutonium. If the most heavily irradiated half of the core fuel-rods were discharged in 1989, they would have contained about 7kg of plutonium, which could have yielded about 5–6kg of separated plutonium if reprocessing losses ranged between 10–30%. In this scenario, North Korea could have loaded a new core or a substantial portion of a new core in 1989 and then falsified the operating records from 1989–92 to make it appear to be the original one.

To add credence to this scenario, US analysts accumulated satellite evidence that North Korea had built, operated and concealed two underground waste storage sites at Yongbyon. One site, originally constructed in the late 1970s, was associated with the IRT-2000 reactor and the radiochemical laboratory. In 1992, as IAEA inspections were underway, satellite imagery indicated that North Korea had covered this old radioactive waste site with earth and had planted vegetation on top of it in order to conceal the location, while building a new waste site nearby. North Korea told the IAEA that this was the original waste site. IAEA inspectors were denied access to the original waste site, which could contain radioactive waste from undeclared plutonium production in the IRT-2000 reactor. Although North Korea admitted in 1993 that it had produced a few hundred milligrams of plutonium in the IRT-2000 reactor in 1975, and claimed that it had mixed waste from these experiments with waste from the 1990 test run at the reprocessing plant, its efforts to conceal the old waste site suggested that additional plutonium had been produced. IAEA analysts estimated that the IRT-2000 reactor could have produced between two and 4kg of plutonium between 1965, when the reactor first became operational, and 1992, when regular IAEA monitoring was authorised.[8] Actual plutonium production is, of course, unknown.

An even more significant waste storage site, known as Building 500, was constructed near the reprocessing plant in 1991. This structure contained a basement that appeared to be divided into compartments for storing liquid and solid radioactive waste. Concrete slabs were laid over the basement and a single-story building was erected on its foundation. In the winter of 1991–92, trenches – presumably to hold piping – were dug between Building 500 and the reprocessing facility, which could have allowed North Korea to pump waste from the reprocessing facility to Building 500 prior to the arrival of IAEA inspectors. If North Korea did carry out substantial undeclared reprocessing and diverted the waste to Building 500, tonnes of radioactive waste would be stored in the basement, which could not be concealed from a close inspection. Measurement of the volume and content of any such waste would provide evidence with which to calculate the extent of undeclared reprocessing and thus the amount of additional plutonium that North Korea could have produced before 1992.

Tipped off by US information, the IAEA sought access to the two suspect waste sites to determine whether radioactive waste, produced by undeclared reprocessing activity, was stored at the sites. In September 1992, IAEA inspectors were allowed to visit Building 500 (which was being used as a military-vehicle repair workshop) but they were told it did not have a basement. After North Korea refused repeated demands for greater access, including for the extraction of samples from underneath the building, the IAEA requested a 'special inspection' of the two suspect waste sites in February 1993. Maintaining that the sites were non-nuclear military facilities beyond the scope of the IAEA inspection mandate, Pyongyang responded, in March 1993, by invoking its right to withdraw from the NPT, setting in motion the 1993–94 nuclear crisis.

Washington's assessment that North Korea might have produced 'enough plutonium for one or possibly two nuclear weapons' or between 8–12kg of separated plutonium before 1992 was based on five factors.

- Firstly, analysis by the IAEA, based on samples from the reprocessing facility, strongly indicated that North Korea had not fully declared its plutonium production prior to 1992, although these sampling 'discrepancies' could not determine how much plutonium North Korea was hiding.
- Secondly, based on satellite surveillance of past operations of the 5MW(e) reactor and estimates of North Korea's de-fuelling capability, the US estimated that most or all of the core could have been unloaded in April–May 1989, containing some 6.5–8.5kg of plutonium after reprocessing. Pyongyang could have falsified the reactor's operating records in order to disguise the insertion of a second fresh core in 1989. North Korea's decision to unload the reactor in June 1994 in a manner that deliberately made it impossible to reconstruct its operating history reinforced suspicion that it wanted to conceal such information.
- Thirdly, North Korea could have employed the small IRT-2000 research reactor to produce small amounts of plutonium every year, perhaps generating between 2–4kg in total.

- Fourthly, two suspect nuclear waste sites provided plausible locations for North Korea to divert and hide substantial quantities of nuclear waste produced as a result of undeclared reprocessing of spent fuel from the 5MW(e) and the IRT-2000 reactors.
- Finally, North Korea's unwillingness to cooperate with the IAEA to resolve discrepancies pertaining to its plutonium declaration and, particularly, its refusal to allow access to the suspect nuclear waste sites – which precipitated a major international crisis – convinced many in Washington that Pyongyang must be hiding something significant.

In short, there was substantial evidence to support the conclusion that North Korea was concealing some plutonium, and plausible scenarios could be constructed to account for enough plutonium for 'one or possibly two' nuclear weapons. The actual amount of plutonium acquired by North Korea before 1992, though, is unknown.[9]

Phase 3 – plutonium programme frozen (1994–2002)
Under the terms of the 1994 Agreed Framework, North Korea agreed to freeze and eventually to dismantle the key facilities associated with its plutonium production programme, including the uranium-conversion and fuel-fabrication plant, the 5MW(e), 50MW(e) and 200MW(e) reactors, and the reprocessing facility. The IRT-2000 reactor (and its related radiochemical laboratory) was exempt from the freeze, on the grounds that it could be used to produce radioisotopes for medical and industrial purposes – subject to IAEA inspections. To verify that the frozen plutonium production facilities were not operating and that all construction had been terminated, the IAEA placed seals on the main access points, installed monitoring devices, and stationed a small team of resident inspectors at Yongbyon, who were allowed to conduct short-notice inspections of different parts of the facilities subject to the freeze.

During the course of the nuclear freeze, the IAEA and North Korea held a series of 'technical discussions', during which Pyongyang agreed to a number of additional containment and surveillance measures to verify the freeze. But Pyongyang resisted any Agency activities that it believed could shed light on its past plutonium production efforts, such as the installation of IAEA monitoring equipment at nuclear-waste tanks, additional sampling at the reprocessing facility or the taking of measurements to determine the plutonium content of the 5MW(e) reactor's spent fuel

'Under the terms of the 1994 Agreed Framework, North Korea agreed to freeze and eventually to dismantle the key facilities associated with its plutonium production programme'

rods.[10] Under IAEA supervision, North Korea also undertook various initiatives to maintain the readiness of its plutonium production facilities. At some establishments, such as the 5MW(e) reactor and the reprocessing plant, the maintenance procedures appeared regular and thorough, while other facilities, such as the 200MW(e) reactor and the fuel-fabrication plant, received little attention.

In addition, the Agreed Framework called on the US to help North Korea to 'stabilise' the 8,000 spent-fuel rods discharged from the 5MW(e) reactor in May–June 1994, pending their eventual removal from North Korea when the first unit of the Light Water Reactor (LWR) project (see below) was completed. After their removal from the reactor, these rods had been stored in a spent-fuel pond next to the reactor building for over two years, during which time a considerable amount of corrosion had occurred. As a result, much of the magnesium cladding and some of the uranium metal had broken loose from the fuel rods themselves, posing a safety risk. Over the course of the next few years, under IAEA monitoring, the spent fuel rods were placed in 400 stainless-steel canisters, each containing approximately 20 rods or fragments. These canisters were filled with inert gas and sealed by US contractors on-site in North Korea, and then placed in underwater racks under IAEA seal. From Washington's perspective, the canning operation helped to prepare the fuel for eventual shipment out of the country because the canisters were designed to be fitted inside shielded shipping casks. From Pyongyang's standpoint, however, canning prevented further corrosion and helped to preserve a reprocessing option if the Agreed Framework failed.

Under the terms of the Agreed Framework, the US, along with Japan and South Korea, formed an international consortium, the Korean Peninsula Energy Development Organization (KEDO), which undertook to provide North Korea with a LWR project, consisting of twin 1,000 MW(e) reactor units, by a 'target date' of 2003. Although construction was delayed for a variety of reasons, site preparation at Sinpo on North Korea's east coast commenced in July 1997, and the concrete foundations of the first reactor unit were laid in August 2002. In accordance with the Agreed Framework, no significant nuclear components were delivered to the LWR project because North Korea was first required to cooperate with the IAEA in accounting for its plutonium production prior to 1992. In any event, the project was formally suspended for one year by KEDO on 4 November 2003, following the collapse of the Agreed Framework. If they were ever to be completed, the LWRs would produce substantial quantities of plutonium in spent fuel, which, in theory, North Korea could extract and divert to weapons production. In practice, North Korea could not acquire spent fuel from the LWR project without detection by

the IAEA, and it has no known facility to separate plutonium from the type of fuel used in the LWRs.

During the nuclear freeze, the US and its allies remained vigilant to the possibility that North Korea might seek to evade the restrictions imposed under the Agreed Framework and continue with its nuclear efforts at clandestine facilities. In early 1998, the US intelligence community concluded, based on data from a variety of sources, including information supplied by defectors and satellite imagery, that North Korea was constructing a large underground facility near Kumchang-ri, about 40km northeast of Yongbyon, which potentially could house a secret reactor and reprocessing facility to produce plutonium. Adding to suspicions, the construction company involved in the Kumchang-ri project was the same one that built the Yongbyon Nuclear Research Centre. Washington's original inclination was to keep the site under observation in order to determine whether North Korea had begun to transport nuclear reactor components to the site, but suspicions about the site were leaked to the press in August 1998.[11]

The suggestion that North Korea might be about to violate the Agreed Framework, combined with the subsequent test of a *Taepo-dong*-1 missile in late August, put enormous political pressure on Washington to address concerns about the construction at Kumchang-ri. After a series of negotiations, North Korea agreed, in March 1999, to allow a US team to visit the Kumchang-ri site in exchange for US food assistance. In May 1999, the US team inspected the site, which consisted of numerous underground tunnels and storage rooms. The inspection revealed that the site was not configured to house an underground reactor and reprocessing facility, much to Washington's embarrassment.

The secret uranium enrichment programme
Under the Agreed Framework, North Korea's capacity to produce additional plutonium at the Yongbyon complex was effectively frozen. As a consequence, its presumed nuclear arsenal was limited to one or two nuclear weapons. Moreover, under the terms of the Agreed Framework, North Korea would eventually be required to account for its undeclared plutonium holdings and to dismantle its plutonium production facilities as a condition for receiving the LWR project. Unless Pyongyang decided to pay the political costs of openly reneging on the provisions of the agreement, it would be forced to give up whatever nuclear weapons capability it had required before 1992. To the extent that maintaining a nuclear hedge was perceived as essential to the survival and defence of the regime, Pyongyang had a strong incentive to develop an alternative means of producing nuclear material, which would allow it ostensibly to comply with the Agreed Framework, while preserving a secret nuclear-weapons programme.

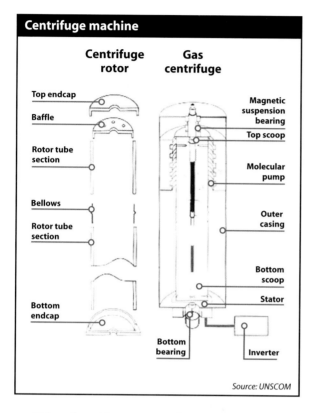

Centrifuge machine

Centrifuge rotor

Top endcap
Baffle
Rotor tube section
Bellows
Rotor tube section
Bottom endcap

Gas centrifuge

Magnetic suspension bearing
Top scoop
Molecular pump
Outer casing
Bottom scoop
Stator
Bottom bearing
Inverter

Source: UNSCOM

a cascade, it is possible to increase the percentage of U-235 from the low level (0.7%) found in nature to higher percentages necessary for nuclear-reactor fuel or nuclear weapons. Generally, enrichment levels of 3–5% U-235 are required to fabricate fuel for the LWRs, while higher levels of U-235 (around 90%) are most desirable for nuclear weapons. A number of countries have built centrifuge plants to produce reactor fuel for civilian nuclear power requirements, including China, Japan and Russia, as well as Uranium Enrichment Services Worldwide (URENCO), a European consortium comprising British, Dutch and German firms.

It is not known exactly what kind of nuclear assistance North Korea received from Pakistan, but it is generally assumed that it could have included technical specifications, sample centrifuge machines, and training that would allow North Korea to duplicate the technology and to assemble a production-scale centrifuge facility. Based on information regarding Pyongyang's procurement efforts, it appears that North Korea is still seeking to obtain components for a type of centrifuge that matches the specifications of one that Pakistan is known to possess.

While working as a metallurgist for a Dutch contractor employed by URENCO in the early 1970s, a Pakistani scientist, A.Q. Khan, obtained the designs for at least three types of centrifuges. The most advanced was the G-2, a German-designed, supercritical centrifuge consisting of two maraging steel rotors connected by flexible bellows, which is capable of rotating at around 500 metres per second. Each machine is about one-metre long with a diameter of 148mm, and is capable of achieving five Separative Work Units (SWUs) per year.

The G-2 eventually became the workhorse of Pakistan's nuclear-weapons programme after Khan returned home to lead its centrifuge programme (as head of Khan Research Laboratories). By the 1990s, Khan was also in charge of Pakistan's efforts to develop a nuclear-capable, long-range, liquid-fuelled missile. Consequently, he became the principal point of contact with regard to nuclear and missile cooperation with North Korea, reportedly making numerous trips to North Korea, beginning in the late 1990s. Because of his role in the development of nuclear weapons, missile warheads and centrifuge technology, it is thought that Pakistani and North Korean technicians may have collaborated on nuclear weapon designs in addition to enrichment and missile technology.

By 2000, US intelligence had begun to detect North Korean attempts to procure equipment and materials that could be used in a centrifuge programme. However, the quantities were small, suggesting a research and development effort, and technical opinion was divided on whether the items were intended for use in centrifuges or for other purposes, such as in missiles. Nonetheless, there was sufficient concern that,

Although neither country has acknowledged the arrangement, it is widely reported that North Korea provided Pakistan with *No-dong* missiles and production technology in exchange for gas centrifuge technology and perhaps other assistance for North Korea's nuclear weapons programme, probably around 1997.[12] For Islamabad, the arrangement was a short cut to acquiring a longer-range missile capable of delivering a nuclear payload, at a time when China was cutting back on its assistance for Pakistan's missile programme. For Pyongyang, the deal provided an attractive alternative technology with which to produce weapons-grade fissile material, in contravention of the terms of the Agreed Framework.

There are many different kinds of centrifuge technology, but the basic principle involves rapidly spinning uranium in gaseous form (uranium hexafluoride) in tubes called rotors. Depending on the design, rotors can be made of high-strength metals (such as certain alloy types of aluminium or steel) or carbon fibre. The centrifugal forces inside the rotor cause a slight separation of lighter U-235 and heavier U-238 atoms and the two 'streams' of uranium hexafluoride are siphoned off under separate withdrawal systems. Each centrifuge machine is capable of a small amount of separation – measured in Separative Work Units (SWUs) – but by passing the slightly enriched stream through an interconnected series of hundreds or thousands of machines, known as

in March 2000, US President Clinton waived a legislative certification that required him to certify that 'North Korea is not seeking to develop or acquire the capability to enrich uranium'. In 2001, a source – said to be a North Korean defector – reported that North Korea had been pursuing a centrifuge enrichment programme for several years, although the location of the production plant and related facilities were apparently not identified.[13] Moreover, North Korea reportedly began seeking large quantities of materials and components that were uniquely associated with centrifuge production, such as high-strength aluminium tubes of specific dimensions and equipment suitable for uranium feed-and-withdrawal systems.[14]

Based on this information, the US intelligence community concluded, in June 2002, that North Korea had embarked on an effort to build a clandestine production-scale centrifuge facility to produce weapons-grade uranium. In October 2002, during his first visit to Pyongyang to open high-level talks between the US and North Korea, US Assistant Secretary of State James Kelly accused North Korean officials of pursing an enrichment programme. According to subsequent US accounts, Vice Foreign Minister Kang Sok Ju, the most senior North Korean official responsible for dealing with the US, 'acknowledged' that the US accusation was accurate. More recently, though, North Korean diplomats claim that Kelly 'misunderstood' Kang, who was merely saying that North Korea had a 'right' to an enrichment programme in view of US 'violations' of the Agreed Framework. In November 2002, the US Central Intelligence Agency (CIA) stated publicly that North Korea is 'constructing a plant that could produce enough weapons-grade uranium for two or more nuclear weapons per year when fully operational – which could be as soon as mid-decade'.[15]

From the available evidence, it appears that North Korea obtained substantial centrifuge assistance from Pakistan and that it is seeking to build a clandestine production-scale centrifuge plant to produce highly enriched uranium for nuclear weapons.

However, the status of North Korea's centrifuge programme, and particularly, how close it is to completion, is very difficult to ascertain, given various uncertainties and Pyongyang's attempts to conceal its activities.

Firstly, it is unclear exactly how much and what type of assistance Pakistan provided. Even if it supplied a full set of machines for a pilot-scale cascade, normally consisting of a few hundred machines, North Korea would need to manufacture or assemble additional machines for a production-scale facility, which typically requires a few thousand centrifuges of the type in Pakistan's possession in order to produce annually enough weapons-grade uranium for a few nuclear weapons. Pakistan denies that it has provided any nuclear assistance to North Korea, although Islamabad is investigating several nuclear scientists who apparently sold centrifuge technology to Iran in 1987.

Secondly, and more important, it is not known whether North Korea has been able to purchase all of the components, materials and equipment necessary to assemble a production-scale centrifuge plant or to what extent it is able to manufacture such items indigenously. There is at least some fragmentary evidence to suggest that North Korea is still shopping around. In April 2003, French and German authorities cooperated to halt a shipment of 22 tonnes of high-strength aluminium tubes, the first instalment of a larger order for 200 tonnes of such tubes. The particular type of aluminium and the dimensions of the tubes closely match the requirements of the rotor casings for the G-2 centrifuge.[16] Allowing for some loss in processing, the 200 tonnes of tubes could be used to manufacture around 3,500 G-2 centrifuges, enough to produce about 75kg of weapons-grade uranium a year or about enough for three nuclear weapons of a first generation uranium-based implosion design, assuming 20–25kg per weapon.[17] Also in April 2003, Japanese authorities, working with officials in Hong Kong, thwarted a North Korean effort to obtain three inverters.[18] These are electronic devices used to generate the direct-current power necessary for the operation of centrifuges, although they also have an application in missile guidance systems. Hundreds of inverters would be required for a production-scale centrifuge plant because each centrifuge machine is run by its own motor.

The interception of the aluminium tubes shipment in April 2003 reinforces the conclusion that North Korea is seeking to build a production-scale centrifuge facility, but these failures in Pyongyang's procurement effort suggest that North Korea may still lack key components. Moreover, aluminium casing tubes are only the 'tip of the iceberg' in relation to the necessary components, materials, and equipment needed to complete a production-scale centrifuge plant. Other critical components, even more difficult to manufacture, include nuclear-grade maraging steel rotors and caps, rotor bearings, and electrical systems. Of course, it is possible that North Korea has been able

'It appears that North Korea obtained substantial centrifuge assistance from Pakistan and that it is seeking to build a clandestine production-scale centrifuge plant to produce highly enriched uranium for nuclear weapons'

to establish a completely undetected procurement apparatus to obtain such items. It seems more likely, though, that North Korea is still in the process of acquiring at least some of the necessary items. As a result, the pace of its enrichment programme could be slowed by stronger interdiction efforts, as envisaged under the new Proliferation Security Initiative (PSI).

Thirdly, the locations of North Korea's centrifuge facility and key ancillary facilities are unknown. Intelligence and military experts believe it is likely that North Korea would choose to build a centrifuge plant underground to guard against detection and military attack. However, sites identified in press reports identifying possible locations for an underground centrifuge plant are mostly based on conjecture, unconfirmed defector reports and analysis of satellite imagery of underground facilities with an unknown purpose.[19] North Korea has myriad military-related underground sites that, in theory, could house a centrifuge plant large enough to produce annually enough weapons-grade uranium for a few nuclear weapons.

In addition to the actual centrifuge facility itself, a North Korean uranium enrichment programme would also require the production of large quantities of uranium hexaflouride (UF_6) feed material, the gaseous form of uranium required for centrifuges. For example, a centrifuge plant consisting of 3,500 G-2-style centrifuge machines would require around 13.5 tonnes of UF_6 feed material per year. North Korea's fuel-fabrication facility at Yongbyong was able to process large amounts of natural uranium yellowcake (U_3O_8) into uranium dioxide (UO_2) and then into uranium tetrafluoride (UF_4), the immediate precursor to the production of UF_6. The fluoride processing lines at the facility are badly corroded, though, and would need to be rebuilt and refitted to resume UF_4 production, perhaps taking a year or so to complete. Of course, North Korea could decide to build a UF_6 feed plant at another location. In the end, production of sufficient UF_6 feed material should not be a major technical hurdle for North Korea's enrichment programme, whether or not such a plant already exists is unknown.

Finally, assuming that North Korea is able to complete a production-scale centrifuge plant, a lengthy period of testing is normally necessary before fullscale sustained production can commence. Centrifuge machines are notoriously temperamental. Operating at high speeds, they can suffer catastrophic failure (known as 'crashing') due to manufacturing for operational errors, requiring that the entire line be shut down in order to replace or bypass the damaged machine. For example, any fluctuation in, or interruption to, the electrical current can prove fatal for centrifuge machines, and North Korea's electrical system is known to be highly unreliable. To overcome this potential problem, North Korea would need to deploy independent generators to produce an uninterrupted power supply in the event of a failure of the national system.

In conclusion, Washington's assessment that a production-scale centrifuge facility could be completed by 'mid decade' is a 'worst case' estimate based on analytical judgements and assumptions, rather than on a wealth of factual information. Fundamentally, the US estimate draws from a sense of how long it would take a country of North Korea's perceived industrial, scientific and engineering potential to complete a production-scale centrifuge facility, assuming that it possessed the necessary technology and that it made a political decision to devote the necessary resources to it. It is possible that North Korea's enrichment programme is even more advanced than US assessments suggest, especially if it has been able to obtain undetected significant quantities of materials and equipment from foreign sources. However, North Korea's enrichment programme probably still faces a number of technical obstacles, which would put the estimated date of completion at the far end of 'mid decade' or even later. In the absence of additional information, it is impossible to make a decisive judgement either way.

Phase 4 – the plutonium programme unfrozen (2002–present)

Following the revelation of North Korea's clandestine enrichment programme and the collapse of the Agreed Framework in late 2002, North Korea disabled IAEA monitoring equipment at the 5MW(e) reactor, the spent-fuel storage pond and the reprocessing facility, expelled IAEA inspectors from the Yongbyon Nuclear Research Centre and took steps to revive its plutonium production programme, which had been suspended since 1994. In the absence of inspectors, it is very difficult to determine the exact status of North Korea's plutonium production programme, although satellite photography and other forms of international monitoring do give some clues. To complicate matters, North Korean officials have made a variety of public and private statements regarding their country's nuclear activities since inspections ended, but these may be for political effect and they cannot be taken at face value. Whatever the uncertainty about the exact status of North Korea's plutonium production

'North Korea's enrichment programme probably still faces a number of technical obstacles, which would put the estimated date of completion at the far end of 'mid decade' or even later'

facilities, enough is known about the technical capabilities of these facilities to produce an informed assessment of North Korea's ability to manufacture additional plutonium in the short term.

Of greatest immediate importance is the fate of the nearly 8,000 irradiated fuel rods discharged from the 5MW(e) reactor in 1994 and stored in an adjacent pond, near to the reprocessing facility. Although the IAEA was not allowed to measure the irradiation levels of the rods, it believes that they contain some 25–30kg of plutonium. Notionally, this is enough for between two and five nuclear weapons, depending on the amount of plutonium lost in the reprocessing process and on the quantity required for each nuclear weapon of North Korean design. If the spent fuel contains 25–30kg of plutonium, the amount actually recovered from reprocessing could be 17.5–27kg, assuming a reprocessing loss of 10–30%. Furthermore, assuming that a first generation implosion design requires 5–8kg of plutonium for each weapon, the separated plutonium would be enough to produce as few as two or as many as five nuclear weapons.

The reprocessing facility was mothballed for nearly eight years, but some maintenance work occurred during the freeze. Most experts believe that North Korea could have restored the facility to operational status within a few months and has likely done so since the inspectors were expelled in December 2002. In theory, the single completed line in the reprocessing plant is capable of reprocessing the 8,000 fuel rods (50 tonnes of uranium) if operations ran continuously for approximately five months, assuming no technical difficulties.

Whether North Korea has reprocessed some or all of the fuel is not known for certain. In early 2003, satellite imagery detected the presence of lorries at the storage site, suggesting that the fuel rods were being removed from the area. In theory, the fuel rods could be moved to the reprocessing facility, or potentially to some other location for processing or protection from military attack. In April 2003, North Korean diplomats told US officials in private that the country had begun reprocessing, and in July they said that reprocessing was completed. In October 2003, North Korea announced publicly that reprocessing of the rods had been completed successfully by the end of June and that the state was using the resulting plutonium to increase its 'nuclear deterrent force'.[20]

The forensic evidence is ambiguous. In June, US monitoring devices located near North Korea reportedly detected slightly elevated levels of Krypton-85 (Kr-85), a radioactive gas released during reprocessing. However, it is extremely difficult to quantify how much spent fuel might have been reprocessed based on these emissions. Although the US has reportedly improved its detection capabilities over the years, evaluation of levels of Kr-85 is complicated because of the presence of background Kr-85 from reprocessing operations in nearby countries, such as China, Japan and Russia, and variations in wind patterns, especially if the amount of fuel reprocessed is relatively small. It is also unknown whether North Korea might attempt to employ technical measures to reduce Kr-85 emissions at the Yongbyon facility, or, indeed, whether there is a second reprocessing facility hidden elsewhere in the country.[21] As of the end of 2003, elevated Kr-85 levels have not been detected since June 2003, and analysts monitoring the Yongbyon reprocessing facility have not observed evidence of continuous operations that would indicate a fullscale reprocessing campaign.

From the available evidence, most government analysts believe that North Korea probably carried out limited reprocessing at the Yongbyon facility in June – perhaps enough to produce one or two nuclear weapons but that Pyongyang stopped short of completing a full campaign. This may have been a test run to appraise the reprocessing facility, or a political tactic to press the US to agree to negotiations. It is possible that North Korea began reprocessing and experienced technical difficulties, or that it exercised caution in response to strong private warnings from Washington that reprocessing would scuttle negotiations. In this case, Pyongyang's public declarations in October that it had completed reprocessing could be intended to strengthen its bargaining position and provide political cover if Pyongyang decides to finish reprocessing at some point in future. An alternative view, is that North Korea's public statements are accurate, and that, in fact, it has completed reprocessing of the available spent fuel, which could not be detected by intelligence means. Once the plutonium is separated, it would be virtually impossible to track and monitor within North Korea using available intelligence resources.

North Korea's ability to produce fresh plutonium in the near term is limited. Since the end of the freeze, it is believed to have refuelled the 5MW(e) reactor at Yongbyon, and restarted it in March 2003 – a view based on the observation of steam plumes from the reactor's cooling tower. The reactor has apparently experienced some start-up problems, which is not surprising after eight years of inactivity, but such difficulties are not

'Once the plutonium is separated, it would be virtually impossible to track and monitor within North Korea using available intelligence resources'

likely to be insurmountable. Assuming maximum power for 300 days, the reactor is capable of producing up to 7.5kg of plutonium per year, perhaps enough for one nuclear weapon, depending on assumptions concerning reprocessing losses and the amount of plutonium required for a nuclear weapon of North Korean design.[22] In its October statement, Pyongyang vowed to reprocess spent fuel from the 5MW(e) reactor as it became available. Plutonium-bearing fuel could be discharged from the 5MW(e) reactor in spring 2004, and assuming some storage time for cooling, reprocessing of the fuel to extract plutonium could be complete by summer or autumn 2004.

In the longer term, North Korea's ability to produce larger amounts of plutonium depends on how quickly it can complete the two larger nuclear reactors that were under construction when the nuclear freeze came into effect in 1994. Estimating the completion time for these larger reactors is difficult. It depends on how far construction had proceded by the time of the freeze, the amount of maintenance work that North Korea has performed in the interim, and the degree of effort and resources that Pyongyang is prepared to invest to finish the projects. It is also unknown whether North Korea has secretly built components or stockpiled materials for the reactors before or during the freeze.

Of the two larger reactors, the 50MW(e) was closest to completion in 1994. At that time, North Korean officials told the IAEA and the US that the reactor was 9–12 months from initial service. However, Pyongyang had an incentive to exaggerate the status of construction because the amount of heavy fuel oil (HFO) delivered to the country under the Agreed Framework was calculated according to the time when the 50MW(e) reactor was expected to be completed. Thus, after one year, the amount of HFO delivered to North Korea under the Agreed Framework was increased from 50,000 tonnes to 500,000 tonnes, the rise roughly representing the expected energy output that North Korea was sacrificing by halting construction of the 50MW(e) reactor.

No technical assessment of the reactor's status in 1994 was conducted. According to IAEA inspectors who visited the reactor throughout the period of the freeze, external work on the main reactor building was complete and the reactor pressure vessel was installed. IAEA inspectors also accounted for slightly more than half of the graphite blocks needed for the reactor core, which were stored in a nearby warehouse, and about one-third of the fuel pieces required for the initial fuel load. These are tagged and sealed under the terms of the Agreed Framework. However, the IAEA was not

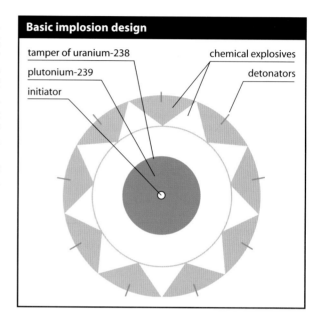

Basic implosion design

tamper of uranium-238

plutonium-239

initiator

chemical explosives

detonators

able to determine the status of several major pieces of equipment and components essential for completing the reactor, such as the fuel-loading machine and the blowers for circulating CO_2 coolant.

Since the end of the freeze in late 2002, there have been no reports of increased activity at the 50MW(e) reactor suggesting that construction has resumed, although some work on essential components could be occurring undetected off-site. Even assuming that key components such as the fuel loading machine and blowers are ready for installation, it would still likely take a few years to complete the reactor. Fuel fabrication, in particular, could pose a delay. Some of the metal parts and equipment used on the process line at the Yongbyon fuel fabrication plant have been badly corroded by fluoride residues because they were not properly cleaned and maintained during the 1994–2002 freeze. Consequently, new equipment would need to be installed before fuel fabrication could resume. Once production recommenced, it would take approximately one year to produce the remaining fuel for the initial core load, unless North Korea has undeclared stocks of fresh fuel or even an undeclared fuel fabrication plant. Finally, once the reactor is complete and fully fuelled, one would normally expect a period of testing before any attempt is made to sustain full-power operations.

Thus, assuming no major technical problems, and that Pyongyang takes the political decision to complete the 50MW(e) reactor, full-power operations could be underway in a few years at the earliest. Resumption of

'Assuming no major technical problems, and that Pyongyang takes the political decision to complete the 50MW(e) reactor, full-power operations could be underway in a few years at the earliest'

large-scale construction at the site and installation of critical equipment and components would probably be detected by satellite, giving advance warning that North Korea was seeking to finish the reactor. In the same timescale that it would take to finalise the 50MW(e) reactor, North Korea could also probably expand its reprocessing capability by completing the second line at its reprocessing facility at Yongbyon. This would provide surplus redundant capacity to process the normal 100-tonne annual spent-fuel discharge from the 50MW(e) reactor. In theory, operating at full power for 300 days a year, the 50MW(e) reactor could produce about 55kg of plutonium per annum, enough for about five to ten nuclear weapons, depending on reprocessing losses and the amount of plutonium required for each weapon of North Korean design.[23] Operating at lower power levels for shorter periods would generate correspondingly less plutonium. Of course, there is no way to know the actual performance capability of the 50MW(e) reactor.

Compared to the 50MW(e) reactor, the much larger 200MW(e) power reactor at Taechon was at an early stage of development in 1994, and has suffered from poor maintenance and exposure to the elements during the period of the Agreed Framework. Indeed, some experts believe that it may be a complete write-off. As far as is known, none of the key components, graphite blocks, or fuel for the reactor has been fabricated. Furthermore, North Korea would need to construct a new reprocessing line to separate plutonium from the reactor's spent fuel. Notionally, the 200MW(e) reactor could produce hundreds of kilograms of plutonium annually, enough for tens of nuclear weapons, but there seems little prospect that it could be completed for many years.

Nuclear weapon design and fabrication
There is virtually no substantial information on North Korean efforts to design and manufacture nuclear weapons, although certain assumptions can be derived from the basic principles that apply to all countries. The common assumption is that North Korea's nuclear weapon design is based on a first generation implosion device, the logical choice for states in the initial stage of nuclear weapon development.

In a first generation implosion device, such as the atomic bomb dropped on Nagasaki in 1945, a solid ball or core of fissile metal (either plutonium or high enriched uranium (HEU)) surrounded by a metal tamper/reflector (usually natural uranium) is compressed by a spherical system of shaped high-explosives, known as a lens. To produce super criticality, a burst of neutrons is introduced at a key instant of compression. The main technical challenge lies in creating a spherical implosion of high-explosives, which requires precise fabrication of the high-explosive lens and exact timing. Failure to achieve this could result in significant loss of nuclear yield or even a dud. Another technical challenge concerns the design of the neutron generator needed to release a burst of neutrons to trigger a chain reaction in the compressed fissile core. The Nagasaki weapon had about 6kg of plutonium in its core, and produced a yield of slightly more than 22 kilotons. Overall, the weapon was some 1.5 metres in diameter, 3.6 metres long, and weighed approximately 4.9 tonnes. Over the years, the advanced nuclear-weapon states have developed a number of different techniques to reduce the amount of plutonium or HEU needed to achieve a desired yield, and to decrease significantly the size and weight of implosion weapons.

An alternative to an implosion design is a gun type device in which a smaller piece of weapons-grade uranium is fired into a larger piece of weapons-grade uranium in order to create a supercritical mass. A chain reaction is initiated with the introduction of a burst of neutrons at a key moment. Unlike an implosion device, which can have either plutonium or uranium as its fissile core, a gun-type device can only be built with HEU because the spontaneous neutrons emitted by plutonium are likely to cause premature criticality (that is, before the nuclear core is fully assembled), significantly reducing the overall explosive yield. Gun-type devices are generally simpler to design and construct than implosion devices, mainly because the high-explosive system used to assemble the critical mass is less complex, but they require considerably more nuclear material to achieve the same yield as is produced by an implosion device. Consequently, implosion designs are generally more attractive to countries with limited amounts of nuclear material. The first gun-type weapon was that dropped on Hiroshima in 1945, which contained about 60 kilograms of HEU and produced a yield of between 12 and 15 kilotons. The overall bomb was 71cm in diameter, 3.4m long, and weighed around 4.3 tonnes.

In estimating the number of nuclear weapons that North Korea might be able to produce, this study assumes that it would require between 5–8kg of weapons-grade plutonium and 20–25kg of HEU for each implosion device, which roughly corresponds to the range of fissile material used by the nuclear-weapon states in their early designs. Using more advanced techniques or aiming to achieve lower yields, nuclear weapons can be built with smaller amounts of plutonium or weapons grade uranium. This study, though, presupposes that North Korea may not have access to such advanced techniques and, therefore, is more likely to pursue simpler and more reliable designs in the range of 10–20 kilotons. However, without knowing the details of North Korea's nuclear-weapon design the actual amount of fissile material used in such a device cannot be determined. The ranges posited in this study cover the most likely possibilities.

Since at least the mid-1980s, North Korea has conducted a series of high-explosive tests, which appear to be related to the development of an implosion system for a nuclear weapon. Prior to 1992, North Korea carried out high-explosive nuclear-related development tests at the Yongbyon Nuclear Research Centre in a nearby stream bed. According to a KGB report of 22 February 1990, leaked to the Russian press in March 1992, the Soviet intelligence agency had already concluded that North Korea had succeeded in developing a 'nuclear explosive device' at the Yongbyon Nuclear Research Centre.[24] The IAEA visited this test site during its various inspections of the Yongbyon establishment in 1992, but it found no evidence of nuclear materials. Later, high-explosive tests were conducted at a nearby site with more sophisticated facilities, known as Youngdoktong. According to a South Korean intelligence report leaked to the country's National Assembly, satellites detected some 70 high-explosive tests at Youngdoktong.[25]

It is difficult to evaluate these North Korean tests with the information available. High-explosive testing for nuclear weapons development has certain distinct characteristics that help distinguish it from high-explosive testing for the development of conventional military ordnance – although any work with shaped charges has some similarities. In theory, using surrogate material for the fissile core, such as natural or depleted uranium metal, such tests can be used to develop an effective nuclear weapon design without the need for a full nuclear test. Whether these tests indicate that North Korea is having difficulty establishing a reliable system, or is seeking to improve on an existing design or to develop new design types, cannot be determined from the available evidence. Given the length of time over which North Korea has apparently conducted nuclear-related high-explosive tests, its ability to manufacture shaped high-explosive charges for conventional munitions, and availability in the public domain of basic information on early implosion designs, the US has believed – since the mid-1990s – that North Korea is capable of designing and building a simple implosion-type nuclear weapon, assuming that it has sufficient stocks of plutonium or highly enriched uranium for such a device. Since North Korea has continued high-explosive testing over the last decade, the current US assessment, that North Korea has built 'simple fission-type' nuclear weapons

without nuclear testing, has become more confident.[26]

If this assessment is correct, a key uncertainty concerns the size and weight of the nuclear weapons, which determine the means of delivery. Clearly, from Pyongyang's standpoint, it would be highly desirable to develop a nuclear weapon small enough and light enough to be delivered by the missiles in its inventory, such as the *No-dong*, which is likely to be more survivable and effective than military aircraft. North Korea's ability to threaten targets beyond the Korean Peninsula, such as Japan or, eventually, the US, would be much more credible if it was able to deliver a nuclear warhead using missiles in its inventory. In this regard, the question of Pakistani assistance is critical. According to press accounts, US intelligence believes that Pakistan may have provided North Korea with nuclear weapon design information and even supplies of high enriched uranium (HEU) under the missile-for-nuclear barter agreement of the late 1990s.[27] With North Korean and Pakistani nuclear and missile personnel apparently working closely together for several years, it is plausible that some discussion of weaponisation would take place. Pakistan's nuclear weapon design, an implosion system utilising HEU instead of plutonium, is based on an early Chinese design, and it is small and light enough to be delivered using the *No-dong* missile – one of the reasons why Pakistan wanted to acquire the *No-dong*. In the worst case, if Islamabad provided North Korea with nuclear weapons design information, it would substantially assist North Korean efforts to develop nuclear weapons that could be delivered by *No-dong* missiles. Even without Pakistani assistance, North Korea may have been able to develop a warhead over the past decade that is small and light enough to be delivered using a *No-dong* missile, but there is insufficient information to make a confident assessment.

In recent years, a few North Korean defectors have come forward claiming that the country possesses nuclear weapons, but much of their information is second-hand and cannot be confirmed. An example of this type of information comes from former Secretary of the North Korean Workers' Party, Hwang Chang Yop, and his assistant, Kim Duk Hong, who defected in 1996. Speaking in 1999, Kim said that Jong Pyong-Ho, a senior party official in charge of military matters, had told Hwang in 1996 that North Korea had five plutonium-based nuclear weapons.[28] Even assuming that Jong actually told Hwang that North Korea possessed nuclear weapons, the possibility remains that the regime would want to create the impression among the party elite that North Korea had a nuclear deterrent

'The US has believed – since the mid-1990s – that North Korea is capable of designing and building a simple implosion-type nuclear weapon'

in order to maintain internal morale. Other defectors have made more extravagant statements, such as a North Korean who said that he was a general in the Korean People's Army and claimed that North Korea had 'dozens' of nuclear weapons.[29] Since the collapse of the freeze, Pyongyang has issued a variety of public and private statements intended to reinforce the impression that North Korea has nuclear weapons, although the regime has not issued a formal public declaration to that effect. The current public formulation is that North Korea possesses a 'nuclear deterrent force'. Of course, Pyongyang has a strong interest in creating the impression that it has nuclear weapons, so its private and public statements are not definitive either way.

In conclusion, the current US assessment that North Korea 'has one or possibly two' nuclear weapons is based on analytical judgements that North Korea has sufficient fissile material and is technically capable of building a simple implosion device, without requiring a full nuclear test, and that Pyongyang has made the political decision to exercise its nuclear option. The original basis for these judgements were developed during the 1993–94 nuclear crisis, and the judgements have become more confident over time. If analysts judged that North Korean scientists and technicians could probably build a first generation device in the early 1990s, it makes even more sense that they could do so a decade later. High-explosive testing has continued during that period, and Pakistani experts may have provided assistance to help North Korea develop a nuclear warhead deliverable by the *No-dong* missile. Whether North Korea actually has nuclear weapons, of course, is not known, but it is impossible to be confident that it does not.

Conclusion

North Korea's current and projected nuclear-weapons capability depends on several key factors. How much nuclear weapons usable material (either separated plutonium or highly enriched uranium) does it possess? How much additional plutonium and HEU will it be able to produce in the future and in what timeframe? What is its capacity to design and fabricate nuclear weapons from its current and projected stocks of nuclear material? In particular, how 'deliverable' would such weapons be?

As with most issues concerning North Korea, there is no definitive answer to any of these questions. The information needed to answer these questions either cannot be obtained or is ambiguous and fragmentary because North Korea has gone to great lengths to

conceal its nuclear activities. Moreover, Pyongyang has actively tried to shape the perceptions of the outside world in one direction or another. In the early 1990s, Pyongyang tried to emphasise the civilian intent of its nuclear programme and sought to downplay the extent of its nuclear capabilities. Since the end of the nuclear freeze in 2002, North Korea has seemingly tried to broadcast its nuclear strength to reassure the party faithful and to deter and intimidate perceived enemies. Whether these statements constitute a boast or a bluff cannot be determined.

During the 1993–94 nuclear crisis, the US assessed that North Korea could have produced enough plutonium prior to 1992 for 'one or possibly two nuclear weapons'. This judgement was based on informed analysis rather than direct empirical evidence of how much plutonium North Korea possessed. That North Korea was pursuing a nuclear-weapons programme throughout the 1980s was clear, despite Pyongyang's belated efforts to justify its programme in terms of nuclear energy production. Moreover, the 'discrepancies' in North Korea's declarations to the IAEA and its refusal to allow access to suspect nuclear waste sites supported the conclusion that it was hiding some plutonium. US experts were able to construct possible scenarios in which North Korea could have manipulated the operations of the Soviet-supplied IRT-2000 reactor and the 5MW(e) reactor to produce something in the range of 8–12kg of plutonium, enough, in theory, for one or possibly two nuclear weapons of the type that North Korea was assumed to be able to manufacture. Given North Korea's long history of nuclear-related high-explosive testing, which began in the mid-1980s, it seemed plausible that North Korea could produce first generation nuclear weapons, assuming that enough plutonium was available.

Throughout the confrontation with the IAEA and the subsequent negotiations with the US over the Agreed Framework, it appeared probable that Pyongyang would not go to such lengths to avoid revealing its plutonium holdings unless it was protecting a quantity of strategic significance – that is, enough for its first nuclear weapon. At the same time, one cannot rule out the possibility that Pyongyang's primary objective was to sustain strategic ambiguity. Whatever its actual plutonium holdings, as long as Pyongyang maintained the perception that it could have enough for one or two nuclear weapons, which was the essence of Washington's public statements during the 1993–94 crisis, it would enjoy some degree of nuclear protection. Over time, the US view that North Korea

'Whether North Korea actually has nuclear weapons, of course, is not known, but it is impossible to be confident that it does not'

had 'one or possibly two nuclear weapons' has grown more confident, although the US does not have proof that North Korea has nuclear weapons.

Whatever its current nuclear inventory, North Korea clearly has the capacity to produce enough fissile material for nuclear weapons in the future – only a few in the near term, but a larger quantity over a number of years. The most immediate threat is from the roughly 8,000 spent fuel rods from the 5MW(e) reactor, which the IAEA estimates contain about 25–30kg of plutonium, enough for between two and five nuclear weapons, assuming a reprocessing loss range of 10–30% and that 5–8kg of plutonium is required for each weapon. The status of this plutonium is unknown. North Korea reportedly began reprocessing this fuel in June 2002, but may not have completed a full campaign, for technical or political factors. North Korea's repeated assertions – first made in private and now publicly – that it has completed reprocessing all of the fuel rods are open to interpretation. Pyongyang may be telling the truth or may be seeking to bolster its perceived nuclear deterrent or it may be trying to create the conditions to finish the job when it believes that the political circumstances are right.

In addition to the plutonium in North Korea's existing stockpile of spent fuel, North Korea has also restarted the 5MW(e) reactor, which, theoretically, can produce enough plutonium for approximately one nuclear weapon per year. However, the reactor may be experiencing some technical difficulties in achieving full power for sustained periods.

Assuming, therefore, one or two nuclear weapons from plutonium separated before 1992, between two and five nuclear weapons from plutonium in North Korea's existing spent fuel, and approximately one additional bomb's worth annually from plutonium produced by the 5MW(e) reactor, North Korea's maximum nuclear arsenal is likely to be limited to 6–12 nuclear weapons over the next several years, if no new facilities to produce plutonium or HEU are completed. This assessment does not include the possibility that North Korea acquired additional nuclear weapons useable material from foreign sources, such as weapons-grade uranium from Pakistan.

In the longer term, North Korea's ability to produce significantly larger quantities of fissile material depends on whether it can complete the 50MW(e) reactor and its presumed centrifuge enrichment plant. It is impossible to predict accurately when these facilities might be completed. From available information, it appears most likely that the 50MW(e) reactor is probably at least a few years from completion and full operation. However, the status of key pieces of equipment are unknown, as is whether the reactor will be able to operate as designed. Presuming that it is completed and operates as designed, the 50MW(e) reactor could produce up to 55kg of plutonium annually, enough for five to ten nuclear weapons, assuming a 10–30% reprocessing loss rate and that between 5–8kg of plutonium is required for each weapon.

Even less is known about the enrichment project. Publicly, the US claims that a production-scale centrifuge facility that is able to produce enough weapons-grade uranium for 'two or more nuclear weapons per year' could be operational as soon as 'mid-decade'. It is not known, though, whether North Korea has been able to obtain the equipment and materials necessary to complete such a facility or the extent to which it can produce such items indigenously. A more conservative estimate is that completion of the plant could be delayed until the end of the decade, especially if interdiction efforts (several of which took place in 2003) can slow the acquisition by North Korea of foreign equipment and materials.

In short, it is impossible to reach a firm conclusion about North Korea's current nuclear weapons capability. On the one hand, a plausible case can be made that North Korea has enough plutonium for a very small number of nuclear weapons, including plutonium that it may have separated before 1992 and plutonium that it may have separated since 2002, and that it is technically able to manufacture a deliverable nuclear weapon from this plutonium. On the other hand, we cannot confirm how much plutonium North Korea has and whether it is able to fabricate a deliverable nuclear weapon from this material. From a public policy standpoint, and given the stakes involved, the case is strong enough that it would be imprudent to conclude that North Korea does not have nuclear weapons.

'Whatever its current nuclear inventory, North Korea clearly has the capacity to produce enough fissile material for nuclear weapons in the future – only a few in the near term, but a larger quantity over a number of years'

North Korea's Chemical and Biological Weapons (CBW) Programmes

Overview

Deciphering the chemical and biological weapons capabilities of any country is a challenge. Chemical weapons (CW) programmes are difficult to trace because many of the facilities potentially involved in military activities are dual-use, with legitimate peaceful purposes, and are relatively easy to conceal. With biological weapons (BW), this is even more the case. With regard to North Korea, assessments are especially difficult due to the fact that – in comparison to other countries suspected of pursuing chemical and biological weapons – the country has remained less accessible in terms of economic and political contacts. Since North Korea is not a party to the Chemical Weapons Convention (CWC), there have never been any official declarations and international inspections of its chemical infrastructure, much less suspect facilities that might be associated with a chemical weapons programme. Also, although North Korea is officially a party to the Biological Weapons Convention (BWC), the Convention lacks a strong verification and inspection mechanism. Another major hindrance to comprehensive insight on North Korea's presumed chemical and biological weapons programmes is that its research and industrial facilities in these areas are relatively isolated from the outside world, so much so that even basic questions of science and infrastructure are uncertain.

In these circumstances, an analysis of North Korea's possible chemical and biological weapons programmes has to rely on public information provided by governments, defectors, and secondary source publications.[1] Such an analysis, made using sources that by their very nature are not comprehensive, will contain many gaps and uncertainties. There are very few details on these suspect programmes that can be specified with confidence. Nonetheless, an analysis based on a variety of sources, particularly official US, Russian and South Korean statements and reports, concludes that North Korea probably has developed chemical weapons to be part of its deployed military capabilities (although there is little authoritative information on the type and amount of agent or delivery means). It is also probable that North Korea has a biological weapons programme at least at the research and development stage. North Korea has dual-use facilities that could be used to produce biological agents as well as a munitions industry that could be used to weaponise such agents. However, there is not enough information to determine whether Pyongyang has progressed beyond the research and development stage for a biological weapons programme and actually possesses stocks of biological weapons.

Chemical weapons programme

Since the early 1990s, official US, Russian and South Korean government publications have all described North Korea as having an active chemical weapons (CW) programme that has gone beyond research and development and includes the actual production and stockpiling of chemical weapons.[2] There is considerable uncertainty, however, over the composition of that stockpile. Given its large – though ageing – chemical industry, North Korea is generally thought to be capable of producing all of the traditional chemical warfare agents (nerve, blister, blood and choking), although it may require imports of some specific precursors to produce nerve agents which are relatively more difficult to fabricate than the first generation blister, blood and choking agents. However, the exact size of the North Korean chemical weapons stockpile remains unknown. Recent South Korean government reports estimate a range of between 2,500–5,000 tonnes, but it is unclear whether these estimates concern the weight of chemical agent or the overall munitions stockpile and even whether they include biological agents. In any event, these figures are highly speculative. There is little authoritative information on the types of chemical munitions that have been stockpiled, but North Korea is capable of using a variety of delivery systems to disseminate chemical agents, including artillery, multiple rocket launchers, mortars, aerial bombs, and missiles, as well as Special Forces. The role of chemical weapons in North Korea's military planning is unknown, but it may be based partially on old Soviet doctrine. US and South Korean forces operate on the assumption that North Korea would use chemical weapons against both military and civilian targets as part of either offensive operations or in retaliation for an attack on North Korea.

Origins and development

In 1954, the North Korean army reportedly established regular chemical and biological defence units, which were most likely modelled on Soviet nuclear, biological, and chemical (NBC) units.[3] According to some press accounts, North Korea's offensive chemical weapons programme also began at this time, relying primarily on assistance from the Soviet Union, but the reliability of these reports cannot be determined.[4] In any event, in the late 1950s, North Korea began to develop an extensive chemical industry.[5] The First Five Year Plan (1957–61)

'An analysis based on a variety of sources, particularly official US, Russian and South Korean statements and reports, concludes that North Korea probably has developed chemical weapons to be part of its deployed military capabilities (although there is little authoritative information on the type and amount of agent or delivery means)'

North Korea's Chemical and Biological Weapons (CBW) Programmes

Major North Korean civilian chemical production facilities	
1. Aoji-ri Chemical Complex	Production of methanol, ammonia, ammonium bicarbonate, coal tar derivatives, liquid fuel products. About 3,500 employees. Annual lignite coal processing capacity of 600,000 tonnes per year. Ammonium bicarbonate production capacity of 100,000 tonnes per year, methane production of 35,000 tonnes per year.
2. April 25th Vinalon Factory	Annual production capacity of 540,000 tonnes of fertiliser, herbicides, and pesticides. Produces civilian products including ammonia, as well as chlorine-based pesticides – probably DDT and chlordane, among others.
3. Chongjin Chemical Fibre Complex	Employs some 3,000 people and has an annual production capacity of 300 tonnes of pesticides, 10,000 tonnes of other chemical products, and 30,000 tonnes of synthetic fibre. Also produces various chemical products, including carbonic acid, formalin and phenol.
4. Chongsu Chemical Complex	Production of large quantities of calcium carbide and smaller amounts of phosphate fertiliser and calcium cyanamide.
5. February 8th Vinalon Complex	One of the largest chemical facilities in North Korea. Employs some 10,000 people, comprises about 50 large buildings, and has an annual production capacity of 50,000 tonnes of vinalon and 10,000 tonnes of movilon. Also produces carbide, methanol, sodium hydroxide, livestock feed, sodium carbonate, vinyl chloride and agricultural insecticide.
6. Hamhung Chemical Factory	Produces civilian chemicals such as sulphuric acid, nitric acid, ammonia and fertiliser products.
7. Hungnam Chemical Fertiliser Complex	Produces civilian chemicals such as ammonium sulphate, ammonium nitrate, phosphate, and urea. Employs more than 10,000 staff and has a production capacity of 1.4 million tonnes (unclear whether annual capacity or other time period).
8. Hwasong Chemical Factory	Produces agricultural chemicals. Annual production capacity of some 2,500 tonnes of phenol. Unknown Iodine capacity.
9. Hyesan Chemical Factory	Produces chemical intermediates such as benzol, phenol and hydrochloric acid.
10. Institute of Chemistry, Hamhung Branch	Research, development, education and training in applied chemistry. Established in 1960 and includes the Revolutionary Historical Relics Preservation Institute, the Institute of Inorganic Chemistry, the Institute of Organic Chemistry, the Institute of Polymer Chemistry, and the Vinalon Institute
11. Manpo Chemical Factory	Produces civilian-related products including ammonia, sodium hydroxide and sulphuric acid.
12. Namhung Youth Chemical Complex	Produces major civilian chemical products including ammonia, ethylene, fertilisers, fibres and paper. Annual chemical production capacity of approximately 550,000 tonnes.
13. Sariwon Potash Fertiliser Complex	Produces fertilisers – planned production target of 510,000 tonnes of potash fertiliser (unclear whether annual production or other time period).

Major North Korean civilian chemical production facilities	
14. Shinhung Chemical Complex	Produces agricultural chemicals and numerous other chemicals such as calcium hypochlorite, caustic soda, dyes, hydrochloric acid, paints, vinyl chloride, polyvinyl chloride, potassium carbonate, sodium carbonate, sodium bicarbonate, barium chloride, ammonium sulphate fertiliser, magnetised fertiliser, slag fertiliser and sulphuric acid fertiliser.
15. Sinuiju Chemical Fibre Complex	Produces calcium cyanamide, chlorine, sodium hydroxide, sulphuric acid, synthetic fibre, paper products and other chemicals. Annual chemical production capacity of 107,000 tonnes.
16. Sunchon Calcium Cyanamide Fertiliser Factory	Produces fertilisers and industrial chemicals such as calcium cyanamide and calcium carbide. Annual chemical production capacity of 100,000–150,000 tonnes. One of North Korea's four major fertiliser plants and is most likely a part of the Sunchon Vinalon Complex.
17. Sunchon Vinalon Complex	North Korea's largest chemical production facility with about 50 affiliated factories. Produces vinalon and other synthetic fabrics, fertilisers, sodium carbonate, vinyl chloride, caustic soda, carbonic acid, livestock feed and methanol. First stage or construction completed in 1989, resulting in an annual production capacity of 50,000 tonnes of vinalon. Final construction reportedly still not completed as of 2000. If the complex is ever finished, its estimated annual capacity will be 100,000 tonnes of vinalon, one million tonnes of carbide, 750,000 tonnes of methanol, and 900,000 tonnes of vinyl chloride.

Based on information from The Nuclear Threat Initiative's website: www.nti.org/e_research/profiles/NK
This draws on information from documents such as 'DPRK Factories Suspected of Producing Chemical Agents', FBIS: KPP20010216000106; 'Alleged Locations of DPRK Nuclear, Biological, Chemical Warfare Facilities Mapped', 6 June 2001, FBIS: KPP20010606000075; 'North Korean Chemical Industry', FBIS: FTS19981230001322; and 'Chemical Engineering, Experts Described', 23 December 1999, FBIS: FTS19991223001168.

placed great emphasis on developing a robust organic and inorganic chemical industry, building on facilities constructed during the Japanese occupation. At the end of 1961, Kim Il Sung issued a 'Declaration of Chemicalisation'. This called for greater efforts to develop various chemical production facilities to support different sectors of the North Korean economy. According to the South Korean Ministry of National Defense, the 1961 declaration reflected North Korean recognition of the importance of chemical warfare.[6] As a result of its large chemical infrastructure, North Korea can produce a number of dual-use chemicals, such as compounds of phosphate, ammonium, fluoride, chloride and sulphur, that could be diverted from civilian chemical uses to support a chemical weapons programme.

By the late 1960s, according to the US Department of Defense, North Korea was believed to have begun experiments with the production of offensive chemical agents.[7] In May 1979, the US Defense Intelligence Agency reported that North Korea possessed only a defensive chemical weapons capability, although it

noted that development of offensive chemical weapons would be the next logical step.[8] Several press reports from the 1980s continued this speculation. The first publicly available official report, to the effect that North Korea had produced chemical weapons agents, was published in January 1987. This publication, by the South Korean Ministry of National Defense, reported that North Korea possessed up to 250 tonnes of chemical weapons – including mustard and nerve agents – designed for delivery by artillery shells.[9]

According to official and secondary reporting, North Korea's chemical weapons arsenal expanded in the early 1990s. However, it is difficult to determine the extent to which such statements reflected actual developments on the ground, or whether they resulted from outside factors affecting public reports of North Korea's programme. Political factors have had an impact. For instance, in 1992, as negotiations for the Chemical Weapons Convention (CWC) were drawing to a close, Seoul sought to publicise the extent of North Korea's chemical weapons programme in a bid to pressure Pyongyang to sign the CWC. In October 1992,

North Korea's Chemical and Biological Weapons (CBW) Programmes

Major North Korean civilian chemical production facilities

EAST ASIA

CHINA

NORTH KOREA

SEA OF JAPAN/
EAST SEA

Korea
Bay

■ Pyongyang

■ Seoul

YELLOW SEA/
WEST SEA

SOUTH KOREA

1 Aoji-ri Chemical Complex
2 April 25th Vinalon Factory
3 Chongsu Chemical Complex
4 Chongjin Chemical Fibre Complex
5 February 8th Vinalon Complex
6 Hamhung Chemical Factory
7 Hungnam Chemical Fertiliser Complex
8 Hwasong Chemical Factory
9 Hyesan Chemical Factory
10 Institute of Chemistry, Hamhung Branch
11 Manpo Chemical Factory
12 Namhung Youth Chemical Complex
13 Sariwon Potash Fertiliser Complex
14 Shinhung Chemical Complex
15 Sinuiju Chemical Fibre Complex
16 Sunchon Calcium Cyanamide Fertiliser Factory
17 Sunchon Vinalon Complex

62 miles

100 km

IISS*maps*

for example, Seoul reported that North Korea had 1,000 tonnes of chemical agent held in six storage facilities, a four-fold increase over the 1987 assessment of 250 tonnes of agent.[10] Pyongyang denied these claims, and countered that the US was storing chemical weapons in South Korea. On 14 January 1993, South Korea signed the CWC when it was opened for signature, and later declared a small stock of chemical weapons, which are being destroyed in accordance with the Convention. North Korea, on the other hand, issued a formal statement on 13 January 1993 denying that it possessed a chemical weapons programme, but it refused to join the CWC.

A second factor, in the mid-1990s, that influenced the public reporting of North Korea's chemical weapons capabilities was the appearance of several prominent defectors, who publicised purported details about North Korea's chemical weapons arsenal, along with related research, production and storage facilities. The most influential of these was Sergeant Yi Chung Kuk, who worked in the Nuclear-Chemical Defence Bureau of the Korean People's Army (KPA) and defected in March 1994. He did so, he said, in order to warn South Korea about the dangers posed by North Korea's chemical weapons programme.[11] Sergeant Yi provided first-hand information on the organisation and equipment of North Korea's chemical defence units, which he was directly involved in, but he also reported secondhand information on offensive chemical weapons activities and facilities. Another key defector was Colonel Choi Ju Hwal, who also worked in the KPA and defected in 1995. Colonel Choi said that he did not have direct knowledge of North Korea's chemical weapons programme, though he claimed to have obtained information from other officials in the Ministry of Defence.[12] Much of Colonel Choi's testimony is identical to information from other defectors, press accounts, and official South Korean government documents, and it is difficult to determine how much is original and how much is derivative. Finally, Hwang Chang Yop, the Secretary of North Korea's Workers Party, defected in August 1996 and said that he had heard from other senior North Korean officials that North Korea had an arsenal of high-grade chemical weapons capable of 'scorching' South Korea and Japan. Mr Hwang did not claim any direct knowledge of chemical weapons production or deployment. Most of the information provided by these North Korean defectors cannot be independently verified, and the usual caveats about information from defectors applies. Nonetheless, their accounts were

widely reported in the South Korean media and may have influenced official assessments by Seoul.

Arguably, Pyongyang had a strong incentive to enhance its chemical weapons programme in the mid-1990s, to compensate for the limits on its nuclear capabilities imposed by the October 1994 Agreed Framework. In addition, the financial limits on modernising its conventional forces may have given Pyongyang more reason to build up its CW capabilities. This speculation cannot be confirmed by direct evidence, but Seoul began to report a greater North Korean chemical weapons capability in the mid-1990s. In 1995, for example, the South Korean Foreign Ministry, the National Unification Board and South Korean military sources reported that North Korea had a stockpile of 1,000–5,000 tonnes of chemical and biological agents, including blister agents, nerve agents, choking agent, blood agent, and tear gas, which could be delivered by artillery, multiple rocket launchers, FROG rockets, and *Scud* missiles.[18] The most recent South Korean Ministry of National Defense report on North Korea's CBW capabilities, from 2001, lists but does not identify by name four research, eight production, and seven storage sites for chemical weapons, and estimates the size of the Pyongyang's stockpile at between 2,500–5,000 tonnes.[19] There is some uncertainty as to whether the various South Korean estimates are for agent or munitions tonnes, and whether they include biological as well as chemical agents.

Official US sources agree on the existence of a North Korean chemical weapons programme, including the stockpiling of agents that could be delivered by a variety of weapons, but Washington has tended to report fewer details than Seoul. In general, US analysts tend to be cautious about the reliability of human information on North Korea's CW programme, and it is extremely difficult to quantify issues concerning potential production rates and possible stockpiles because North Korean chemical facilities are not subject to international inspections, and satellite intelligence has little value in distinguishing between chemical production for military or civilian purposes. A 2001 US Department of Defense report identifies nerve, blister, blood, choking and tear gases as among the agents the North Koreans can produce and assesses that North Korea possesses a 'sizeable stockpile' of these agents, without estimating a specific quantity of agent.[20] According to the US, there may be limits on the North's production capacity. For example, the senior US military official in Seoul, General Schwartz, has testified that the North is capable of independently producing

'Arguably, Pyongyang had a strong incentive to enhance its chemical weapons programme in the mid-1990s, to compensate for the limits on its nuclear capabilities imposed by the October 1994 Agreed Framework'

North Korea's Chemical and Biological Weapons (CBW) Programmes

Defector reporting on North Korea's chemical weapons programme		
NAME	**BACKGROUND**	**DEFECTOR COMMENT**
Yi Chung Kuk	Sergeant in the 18th Nuclear and Chemical Defence Battalion in the early 1990s. Defected in March 1994.	Warned that North Korea was capable of killing all people in South Korea with the chemical and bacterial weapons it possessed; identified various institutes and facilities associated with research, production, storage, and testing of biochemical weapons.[13] He provided detailed information on the 18th Nuclear and Chemical Defence Battalion, which is divided into 6 separate companies, responsible for nuclear/chemical detection, reconnaissance, and decontamination. He also provided information on Factory No. 279 and the No. 398 Research Institute in Sokam-ri, said to be responsible for development and production of chemical defence equipment. He also linked the Sunchon Vinalon Complex to North Korea's chemical weapons programme.
Choi Ju Hwal	Served in the Ministry of the Defence from 1968 to 1995. Defected from a post of Colonel and Chief of joint venture section of Yung-Seong Trading Company in 1995. (Choi acknowledged that he did not have direct knowledge of CBW programmes, but says he obtained second-hand information from other officials.)	In 1997, Choi said that North Korea has stored over 5,000 tonnes of toxic gases, including nerve gases (sarin, soman, tabun, and V agents), first-generation blister gases (lewisite and mustard gas) and blood agents (hydrogen cyanide and cyanogen chloride).[14] Choi identified numerous facilities associated with CW research and production, including several civilian chemical factories involved in vinalon production. He also said that Major Kim Jong Chan, who was an assistant military attaché at the North Korean Embassy in East Germany in the late 1970s, obtained technical knowledge on manufacturing 'extremely poisonous gases' and that North Korea started to manufacture new types of 'poisonous gases' in the mid–1980s.
Yi Sun Ok	Inmate at a North Korean prison. Defected in 1995.	Ms Yi said that some 150 fellow inmates died due to a chemical weapons test in 1990.[15]
Hwang Chang Yop	Secretary of North Korea's Workers Party. Defected in August 1996.	In August 1996, Hwang claimed that North Korea had both nuclear and chemical armed missiles capable of 'scorching' South Korea and Japan.[16] He quoted the North Korean leadership as saying that North Korea ranked third or fourth in the world in chemical weapons.
Yi Chun Son	Served as a commander at a 'missile station'. Defected in 1999 from the North Korean Ministry of People's Armed Forces.	Yi said that chemical agents are produced in Factory 102 in a process involving nitrogen dioxide mixed with sulphur dioxide which is then heated and combusted with mercury. The product is inserted into glass bottles, which are taken by helicopter to the No.108 Factory in Kanggye (which produces artillery shells). The chemical process described by Yi does not correspond to any known production process for a chemical weapon agent.
Yi Mi (pseudonym)	Worked at the Yongbyon nuclear complex. Defected in September 2000.	Yi said that the 304 Lab mainly worked on nuclear weapons development but also conducted research and development in chemical weapons.[17]

NOTE: No effort has been made to verify the claimed backgrounds and information provided by these individuals. The chart is intended to illustrate the range and type of raw information provided by defectors.

Common chemical weapons agents			
AGENT	AGENT IDENTIFICATION	PERSISTENCY	MAJOR PROMPT EFFECTS
Blister agents			
Mustard	H	Very High	Skin blistering, conjunctivitis, damage to airways, death.
Lewisite	L	Moderate	Cutaneous (skin): Pain and irritation of eyes and skin followed by blisters and lesions on the skin. Pulmonary (inhalation): runny nose, hoarseness, bloody nose, sinus pain, cough. Intenstinal: diarrhoea, nausea, vomiting.
Choking agents			
Phosgene	CG	Low	Coughing, blurred vision, shortness of breath, nausea, pulmonary oedema, heart failure, death.
Diphosgene	DP	Low	Coughing, blurred vision, shortness of breath, nausea, pulmonary oedema, heart failure, death.
Vomiting agents			
Adamsite	DM	Low	Coughing, severe headache, muscle spasms, chest pains, shortness of breath, nausea, vomiting.
Blood agents			
Cyanide (hydrogen cyanide and cyanogen chloride)	ANCK	Low	Rapid breathing, dizziness, weakness, headache, nausea, vomiting.
Nerve agents			
VX	VX	Very High	Salivation, runny nose, sweating, shortness of breath, leading to muscle spasms, unconsciousness, death.
Sarin	GB	Low	Salivation, runny nose, sweating, shortness of breath, leading to muscle spasms, unconsciousness, death.
Tabun	GA	Moderate	Salivation, runny nose, sweating, shortness of breath, leading to muscle spasms, unconsciousness, death.
Soman	GD	Moderate	Runny nose, watery eyes, rapid breathing, nausea, leading to unconsciousness, paralysis, respiratory failure, death.

For further information see:
- Organization for the Prohibition of Chemical Weapons (OPCW): www.opcw.org/resp/html/cwagents.html
- World Health Organisation (WHO): www.who.int/csr/delibepidemics/biochem_threats.pdf
- Carnegie Endowment for International Peace: www.ceip.org/files/publications/RegimeAppendix7.asp?p=
- NATO Handbook on the Medical Aspects of NBC Defensive Operations AmedP-6(B): www.fas.org/nuke/guide/usa/doctrine/dod/fm8-9/toc.htm
- US Government, the Chemical & Biological Warfare Threat; US Army Medical Research Institute of Chemical Defence, Chemical Casualty Care Division, http://ccc.apgea.army.mil

North Korea's Chemical and Biological Weapons (CBW) Programmes

components only for first generation (i.e. World War I-type) chemical agents (e.g. phosgene and mustard).[21] Imports of some precursors may be necessary for the production of more advanced nerve agents. Official US sources agree with South Korean reports that North Korea has weaponised chemical weapons agents for deliver by artillery, missiles, and aircraft, as well as unconventional means, but US public reports generally do not discuss suspect or possible research, production, and storage sites associated with chemical weapons.

North Korean defectors and various secondary sources have provided detailed information about facilities purportedly involved in research, production, and storage of chemical precursors, agents and munitions.[22] According to these sources, North Korea's chemical weapons stockpile includes first generation blister agents (lewisite and mustard), various nerve agents (sarin, soman, tabun, and V-agents), and blood agents (hydrogen cyanide and cyanogen chloride). Chemical weapons research is said to take place at various universities and at a number of institutes under the aegis of the Second Natural Science Academy. Chemical weapons production facilities are reported to include the Kanggye Chemical Factory and Factory No. 108 in Chagang Province, the Sakchu Chemical Factory in North Pyongan Province, the Ilyong Branch of the Sunchon Vinalon Factory in South Pyongan Province and Factory No. 297 in Pyongwon, South Pyongan Province.

In addition, a number of civilian chemical facilities have been implicated in chemical weapons production, such as the Manpo Chemical Factory and Aoji-ri Chemical Complex. Defectors and press stories also report that chemical agent storage sites are located in the cities of Masan-dong, Samsan-dong, and Sariwon, and in the greater Pyongyang area. These facilities are reportedly comprised of storage tanks housed in warehouses and buildings above ground, partially buried structures, and underground tunnels. It is alleged that chemical weapons agents are transferred to facilities at Sakchu or Kanggye for loading into munitions, which include 80mm artillery shells, 240mm rockets, aerial bombs, and aerial spray tanks. Following final assembly and filling, chemical munitions are reportedly stored at the Maram Materials Corporation and the Chiha-ri Chemical Corporation, located in Masan-dong, Pyongyang, and Anbyon, Kangwon Province, respectively. Most of this information cannot be independently confirmed.

Potential military uses for chemical weapons

Assuming that North Korea maintains a stockpile, chemical weapons agents and munitions could play a role in complementing Pyongyang's conventional military power in offensive or defensive operations. In theory, North Korean forces could use chemical weapons against US and South Korean forces to reduce these forces' combat effectiveness, deny the use of mobilisation centres, storage areas, and military bases, and hinder the arrival of reinforcements from overseas. Non-persistent chemical agents could be used to help break through defensive lines or to hinder an allied counterattack. Persistent chemical agents could be used against fixed targets, including command and control centres, logistics hubs, and airbases. North Korean forces appear to be prepared for operations in a contaminated environment. Chemical defence battalions are reportedly integrated into larger ground force units, and many troops are reportedly equipped with chemical protection equipment, including masks, suits, detectors and decontamination systems. North Korean troops are also said to participate in chemical exercises in an attempt to develop mission capability under chemical warfare conditions.

Of course, these defensive measures could reflect North Korean expectations that their forces may be subjected to a chemical attacks. Nonetheless, US and South Korean military commanders assume that North Korean offensive military plans include the use of chemical agents delivered by a variety of traditional means, such as ballistic missiles, artillery rockets and shells, mortars, and aerial bombs and sprays, against both military and civilian targets. Delivery by Special Forces is also a possibility. Aside from their potential role in offensive operations, chemical weapons presumably contribute to North Korea's deterrent posture, especially since North Korea's conventional capabilities have eroded relative to US and South Korean forces. Although Pyongyang officially denies that it possesses chemical weapons, the widespread belief that North Korea has a substantial chemical weapons arsenal – noted in official US and South Korean government reports – only serves to reinforce the view in the US, South Korea and Japan that a conflict on the Korean Peninsula would result in the use of chemical weapons against civilian and military targets.

Biological weapons programme

There is less public information on North Korea's

'US and South Korean military commanders assume that North Korean offensive military plans include the use of chemical agents delivered by a variety of traditional means, such as ballistic missiles, artillery rockets and shells, mortars, and aerial bombs and sprays, against both military and civilian targets'

Common biological weapons agents

TYPE	SYMPTOMS	CHARACTERISTICS
Bacteria		
Bacillus anthracis (anthrax)	Pulmonary (inhalation): difficulty breathing, exhaustion, toxemia, terminal shock. Cutaneous (skin): itching, small lesions and possible blood poisoning. Intestinal: nausea, fever, diarrhoea.	Mortality (if untreated): Pulmonary 80–95%. Cutaneous 5–20%. Intestinal 25–60%. Incubation period: Symptoms usually occur within 7 days. Not contagious.
Vibrio cholerae (cholera)	Diarrhoea, vomiting and leg cramps. Rapid loss of body fluids, dehydration and shock.	Mortality (if untreated): 5–10%. Death in 1–3 hours. Not contagious.
Yersinia pestis (plague)	Fever, headache, extreme exhaustion, development of painful, swollen lymph nodes – called buboes. Leads to blood infection and pneumonia.	Mortality (if untreated): 50–60%. Incubation period: 1–3 days. Death in 2–6 days. Contagious through respiratory droplets from pneumonia patients.
Salmonella Typhi (typhoid fever)	Sustained fever, malaise, chills, stomach pains, headache, loss of appetite, occasional rash of flat, rose-coloured spots.	Mortality (if untreated): 12–30%.
Typhus	Fever, headache, chills, and general pains caused by a whole body rash.	Mortality (if untreated): 30–50%. Incubation period: 6–12 days. Not contagious.
Mycobacterium tuberculosis (tuberculosis)	Coughing, pain in the chest, fatigue, weight loss, loss of appetite, chills, fever, sweating at night, coughing up blood or sputum.	Mortality (if untreated): 30–50%. Incubation period: 14 days–1 year. Contagious.
Virus		
Haemorrhagic fever (Korean Strain)	Fever, fatigue, dizziness, muscle aches, exhaustion. Can cause bleeding under the skin, in internal organs, or from body orifices; coma, delirium, and seizures.	Mortality (if untreated): 5–15%. Incubation period 2–21 days. Contagious.
Variola (smallpox)	First symptoms include fever, malaise, aches, leading to high fever, rash, and crusting scabs.	Mortality (if untreated): 30–40%. Incubation period: 7 to 17 days. Contagious.
Yellow Fever	High fever, chills, headache, muscle aches, vomiting. Can lead to shock, kidney and liver failure (causing jaundice).	Mortality (if untreated): 5–40%. Incubation period: 3–6 days. Not Contagious.
Toxin		
Clostridium Botulinum (Botulinum toxin; botulism)	Nausea, weakness, vomiting, respiratory paralysis.	Mortality (if untreated): 60–90%. Incubation period: 12–36 hours after inhalation. Death in 24–72 hours, illness for months if not lethal. Not contagious.

For further information see:
- World Health Organisation (WHO): www.who.int/csr/delibepidemics/en/annex3May03.pdf
- NATO Handbook on the Medical Aspects of NBC Defensive Operations AmedP-6(B): www.fas.org/nuke/guide/usa/doctrine/dod/fm8-9/2toc.htm
- US Army Medical Research Institute of Infectious Diseases, USAMRIID's Medical Management of Biological Casualties Handbook: www.usamriid.army.mil/education/bluebook.html
- Centres for Disease Control: www.cdc.gov

North Korea's Chemical and Biological Weapons (CBW) Programmes

Major North Korean civilian biotechnology facilities	
1. Aeguk Compound Microbe Center	Research, development and production of microbial-based fertiliser supplements. Supplies microbial stock to branch compound microbial fertiliser factories, of which there are reportedly over 120 in North Korea.
2. Aeguk Preventative Medicine Production Factory	Comprises 10 laboratories and various workshops devoted to research, development, and production of vaccines and medicines. The main product has been hepatitis B vaccine produced through the use of recombinant yeast.
3. Branch Academy of Cell and Gene Engineering	One of nine research branches of the Academy of Sciences. Conducts research on cellular biology and genetic engineering. Reportedly involved in the manufacture of human growth hormone, restriction endonucleases (enzymes used in genetic recombination techniques), snake-venom derived anticoagulant, genetically modified crops, and vaccine research.
4. National Sanitary and Anti-Epidemic Research Centre	The Haemorrhagic Fever Laboratory is established under this facility. The Institute's main duties include support for national sanitation, providing inoculations against various diseases, and administering quarantines.
5. Endocrinology Institute	Mainly diagnoses and treats diabetes and various diseases. Reportedly has three laboratories: the Biochemistry Laboratory; the Experimental Treatment Laboratory; and the Generic Engineering Laboratory.
6. Industrial Microbiology Institute	Research, development and production of microbial cultures with applications in areas such as feed supplements, medicines and vaccines, fermentative industries, and food and beverages.
7. Munchon Agar Plant	Agar (growth media) production. As of 1992, the annual agar production capacity of the factory was 200 tonnes.
8. Pharmaceutical Institute of the Academy of Medical Sciences	Research and development of medicaments. Reported work includes the development of nutritive supplements, including production of amino acids derived from industrial by-products. Described as being located at or near the logistical chain of medical supply organisations, it is likely that the Institute is in Pyongyang.
9. Pyongyang Pharmaceutical Factory	The United Nations Children's Fund (UNICEF), in collaboration with Diakonie Emergency Aid (DEA) of Germany, has provided support to upgrade North Korea's main pharmaceutical factory. As of August 2000, the factory produced seven drugs, including antibiotics and multivitamins, with raw materials provided by DEA, according to UNICEF. Support for capacity development in Good Manufacturing Practices (GMP) has been provided through in-country training.
10. Synthetic Pharmaceutical Division, Hamhung Clinical Medicine Institute	Research and development of medicaments and clinical diagnostics. Work includes improving diagnostics testing for antibiotic susceptibility by bacteria. Researchers also surveyed folk remedies and medicinal herbs for inclusion in pharmaceutical development.
11. Taedonggang Reagent Company	Research and Development of vaccines (e.g. hepatitis B), development and production of diagnostic equipment, electron microscopes, and reagents. Developed electron microscopes, which have been advertised for sale abroad. Previously known as the November 19 Institute.

Sources: Nuclear Threat Initiative: www.nti.org; 'DPRK's NAS Pursues Cultivation of Stock Bacteria for Microbial Fertilizers', *Chungang Ilbo*, 17 January 2000; UNICEF Emergency Programmes: DPRK Korea Donor Update, 7 Aug 2000: www.reliefweb.int

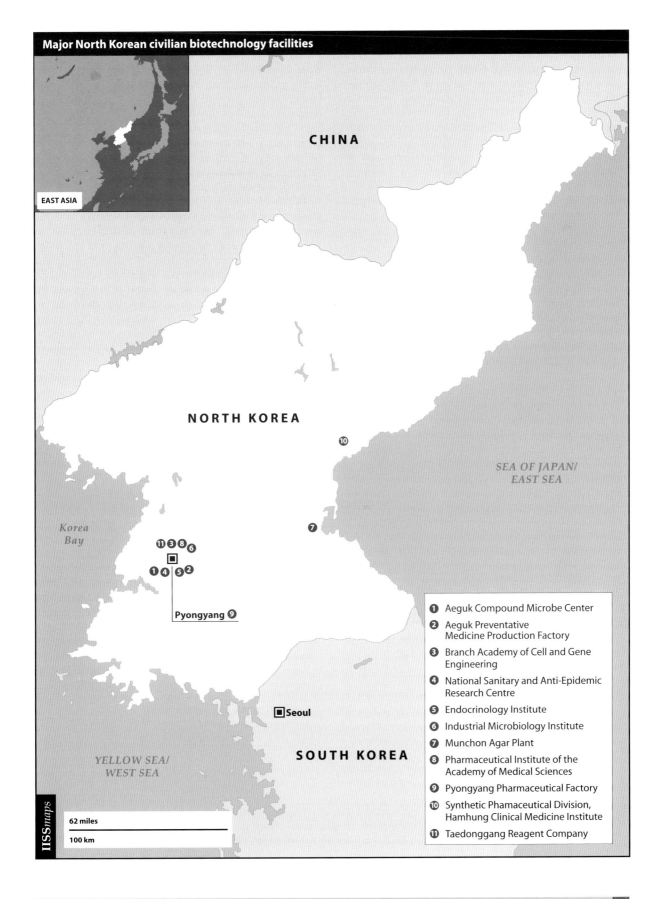

Major North Korean civilian biotechnology facilities

EAST ASIA

CHINA

NORTH KOREA

SEA OF JAPAN/
EAST SEA

Korea
Bay

Pyongyang

Seoul

YELLOW SEA/
WEST SEA

SOUTH KOREA

62 miles

100 km

1 Aeguk Compound Microbe Center

2 Aeguk Preventative
   Medicine Production Factory

3 Branch Academy of Cell and Gene
   Engineering

4 National Sanitary and Anti-Epidemic
   Research Centre

5 Endocrinology Institute

6 Industrial Microbiology Institute

7 Munchon Agar Plant

8 Pharmaceutical Institute of the
   Academy of Medical Sciences

9 Pyongyang Pharmaceutical Factory

10 Synthetic Phamaceutical Division,
    Hamhung Clinical Medicine Institute

11 Taedonggang Reagent Company

biological weapons programme than on its chemical weapons programme. Official US, Russian and South Korean reports agree that North Korea has conducted biological weapons research, but there is considerable uncertainty as to whether Pyongyang possesses biological weapons and, if so, the types of agents involved. While official South Korean sources claim that North Korea has weaponised one or two biological agents, official US and Russian sources characterise North Korea as 'capable' of producing a variety of agents, including anthrax, cholera and plague without judging that North Korea has actually produced biological weapons. Given the dearth of information, it is impossible to make a firm judgement either way. Various defectors and press reports give details of biological weapons research, testing and production, but such information cannot be confirmed. There is no authoritative information on the potential role of biological weapons in North Korean military strategy, beyond speculation that biological weapons may be relatively less significant than chemical weapons, which have more utility as a battlefield weapon, and nuclear weapons, which are a more capable mass destruction weapon.[23]

Virtually nothing is known about the history of North Korea's biological weapons programme. Official US sources state that North Korea has pursued a biological warfare capability since the 1960s.[24] During this time, according to press reports, a laboratory was established under the authority of the Academy of National Defence and 10–13 different pathogens were investigated, including anthrax, cholera, bubonic plague, smallpox and yellow fever, some of which reportedly were imported from culture collections in Japan.[25] According to another secondary source, construction of an underground biological weapons research and development facility was completed in the 1970s[26]. This facility was located in Onjong-ri, South Pyongan Province and conducted research, development, and testing of biological weapons agents on small laboratory animals.

A 1998 White Paper released by the South Korean Ministry of National Defense, reported that, 'by 1980, [North Korea] had succeeded in its experiments in bacteria and virus cultivation to produce biological weapons, and by the late 1980s had completed live experiments with such weapons.'[27] This is generally consistent with a 1993 report by the Russian intelligence service on proliferation, which stated that North Korea was performing 'applied military-biological research' with anthrax, cholera, bubonic plague and smallpox at a number of institutes and universities and testing biological weapons on North Korean islands.[28] South Korean press and other unofficial sources go even further, claiming that, in the early 1980s, North Korea began actual production of biological agents and obtained a turnkey plant for agar (growth media) from East Germany in 1984 to further the biological weapons programme.[29] In contrast, a 1997 US Department of Defense report judged that North Korea's biological weapons programme was probably still at the level of research and development.[30]

Whatever the status of its biological weapons efforts, North Korea has developed a number of dual-use biotechnology facilities that could be used to research biological weapons agents and produce militarily significant quantities of biological agents. But this infrastructure is not highly developed and there is no definitive evidence that it is being used for this purpose.[31] North Korea joined the BWC on 13 March 1987 (followed by South Korea on 25 June 1987), but the convention has no provisions for mandatory declarations or inspections of civilian or suspect military biological facilities.

The most recent official US and South Korean reports agree that North Korea has a biological weapons programme, although only Seoul reports that it has advanced beyond the research and development stage. In 2001, for example, a South Korean defence White Paper described the North Korean threat as including 'chemical and biological weapons such as anthrax of which North Korea is believed to hold a stockpile of 2,500–5,000 tons.'[32] The report does not address the issue of delivery systems, other than to note that North Korean Special Forces could launch attacks with biological weapons. Another South Korean Ministry of National Defense report from 2001 claims that North Korea possesses three research and six production facilities to support its biological weapons programme and has weaponised one or two types of biological agents.[33] In contrast, the most recent public US government report, from 2001, says that 'North Korea is believed to possess a munitions-production infrastructure that would allow it to weaponize biological warfare agents, and may have biological weapons available for use'.[34] According to press accounts, the US intelligence community has assessed with 'medium' confidence that North Korea possesses stocks of smallpox virus, but the evidence is not definitive.[35]

Most of the detailed information about North Korea's biological weapons programme has come from defectors and other secondary sources of unknown

'Official US, Russian and South Korean reports agree that North Korea has conducted biological weapons research, but there is considerable uncertainty as to whether Pyongyang possesses biological weapons and, if so, the types of agents involved' reliability. According to Choi Ju Hwal, the Germ Research Institute in the General Logistic Bureau of the Armed Forces Ministry is responsible for developing biological weapons.[36] Yi Chung Kuk, meanwhile, claims that biological weapons research and development is carried out at the Microbiological Institute and that there are other facilities in North Korea for producing and storing biological weapons.[37] Yi Sun Ok, who was an inmate at a North Korean prison camp, claims she witnessed biological weapons experiments in mid-1980s, which resulted in the deaths of some 50 inmates.[38] However, none of these reports can be confirmed.

A number of secondary sources provide additional details on facilities and suspected agents said to be involved in North Korea's biological weapons programme. According to one report, research on anthrax, bubonic plague, smallpox, yellow fever, cholera and other pathogens is carried out at the National Defence Research Institute and Medical Academy (NDRIMA).[39] Another report says that North Korea's inventory of biological agents includes anthrax, botulism, cholera, haemorrhagic fever (Korean strain), bubonic plague, smallpox, tuberculosis, typhoid, typhus, and yellow fever.[40] Another claims that 13 types of biological weapons agents are produced at the Workers Party's Central Biology Research Institute, the Preventive Military Medical Unit, and the February 25th Plant in Chongju, North Pyongan Province.[41] But these reports also cannot be confirmed. To date there is no reliable information available to confirm whether North Korea has engaged in the development of genetically modified biological agents.[42]

In conclusion, there is not enough information to reach a firm judgement on the progress of, or possible effectiveness of, North Korea's biological weapons programme. This is understandable, given North Korean secrecy and the inherent difficulties of detecting and assessing biological weapons programmes, compared to nuclear or even chemical weapons activities. US, South Korean, and Russian official sources agree that North Korea has conducted research on a variety of biological agents, but only Seoul reports that North Korea has actually produced stocks of one or two types of biological weapons. The basis for this assessment is unspecified. Given its biotechnical infrastructure, North Korea is capable of producing significant amounts of common biological agents, such as anthrax, and delivering these agents through a variety of conventional and unconventional means, but it is not known how important Pyongyang views the

development and deployment of a biological weapons capability. In any event, the possibility that North Korea may have biological weapons contributes to deterrence.

Conclusion

The available evidence suggests that North Korea probably possesses both a chemical and biological weapons programme, although they may differ in terms of scope and state of advancement. The chemical weapons programme probably involves some chemical weapons production and stockpiling, although the amount and types of agents that have been produced, the number and types of munitions that have been stockpiled, and the location of key research, production, and storage facilities cannot be assessed with high confidence. North Korea is thought to be capable of producing a variety of traditional blister, blood, choking and nerve agents, although there may be limits on what it can produce in its ageing chemical industry. Meanwhile, given its munitions industry, North Korea is thought capable of producing a variety of delivery systems for chemical weapons, including artillery, multiple rocket launchers, mortars, aerial bombs, and missiles. The extent to which Pyongyang has chosen to deploy these capabilities is unknown, but US and South Korean forces prudently assume that North Korea possesses chemical weapons and is prepared to use them against military and civilian targets in offensive operations or in retaliation for an attack on North Korea. By comparison, less is known about North Korea's presumed biological weapons programme. While there is general agreement that North Korea has conducted research and development on biological agents, there is not enough information to conclude whether it has progressed to the level of agent production and weaponisation, although North Korea is most likely technically capable of both.

Whatever the actual status of North Korea's chemical and biological capabilities, the perception that it has, or likely has, chemical and biological weapons contributes to Pyongyang's interest in creating uncertainties in Washington, Seoul and Tokyo and raises the stakes to deter or intimidate potential enemies. From Pyongyang's perspective, chemical and biological weapons could have utility both on the battlefield and at the strategic level. US and South Korean military commands have to operate on the assumption that North Korea maintains a large stockpile of chemical and possibly biological munitions integrated with its conventional forces and deployed for use on the battlefield. This complicates allied

'Whatever the actual status of North Korea's chemical and biological capabilities, the perception that it has, or likely has, chemical and biological weapons contributes to Pyongyang's interest in creating uncertainties in Washington, Seoul and Tokyo and raises the stakes to deter potential enemies'

North Korea's Chemical and Biological Weapons (CBW) Programmes

military planning for defence against any North Korean attack or for conducting offensive operations against the North. Some measures have been taken to strengthen allied troops' CBW defences, but it is difficult to accurately assess their effectiveness without knowing the size, composition, or delivery means of North Korea's presumed chemical weapons arsenal. At the strategic level, the potential delivery of large quantities of chemical or biological agents to nearby targets (such as Seoul) and smaller quantities to more distant targets (such as Tokyo) could cause significant civilian casualties, depending on the amount and type of agent, the delivery means, the extent of civilian defence measures, and many other factors. In any event, the plausible threat that North Korea might use chemical or biological weapons, if the survival of the regime was at stake, contributes to deterrence and discourages Seoul and Tokyo from pursuing policies that could increase the risk of conflict and drive Pyongyang to take desperate measures.

North Korea's Ballistic Missile Programme

Overview

North Korea's interest in developing a ballistic missile capability appears to stem from its continuing efforts to establish and maintain robust military forces against South Korea, Japan and US forces in the region. As such, the reach of North Korea's missile programme has expanded from, in the 1960s and 1970s developing and deploying tactical artillery rockets, to developing and deploying short-range ballistic missiles in the 1980s and, in the 1990s, developing and deploying medium-range ballistic missiles. Systems capable of greater ranges are currently under research and development.

North Korea's missile programme is based primarily on Soviet *Scud* missile technology, and Pyongyang is believed to have developed an infrastructure for missile research and development, testing, and production. This indigenous infrastructure has been supplemented by imports of specialised material and components. Although the primary motivation for North Korea's missile programme appears to be related to security concerns, the export of missiles and missile-related technology to numerous customers in the Middle East and South Asia for cash and barter has become an important secondary consideration.

Currently, North Korea has produced and deployed short-range *Hwasong*-5/-6 missiles (*Scud*-B/-C types), which can reach targets throughout South Korea, and medium-range *No-dong* missiles, which can reach targets across Japan. The exact size, disposition, and armament of these North Korean missile forces are unknown, although they are presumed to be modelled on Soviet doctrine, and organised into launch battalions with four to six mobile launchers per battalion plus support vehicles. (Normally, four missiles would be deployed with each launcher, one missile on the launcher and three on a reload missile carrier.) A number of different bunkers, shelters and tunnels associated with missile forces have been identified throughout North Korea. Conservatively, North Korea is estimated to deploy around five to seven *Hwasong*-5/-6 battalions and between two and three *No-dong* battalions – a capability, including reserves, equal to several hundred missiles. Armed with high-explosive warheads, North Korea's missiles could serve as terror weapons against foreign cities or as long-range, though inaccurate, artillery against military targets. US and allied military commands assume that North Korea has armed some missiles with chemical and possibly biological warfare (CBW) payloads. Pyongyang certainly has this capability, but amounts and types of possible CBW

agents cannot be determined. The *No-dong* missile is capable of accommodating a simple fission-type nuclear weapon, if Pyongyang has produced such a warhead.

In comparison to its deployed single-stage short- and medium-range missiles, even less is known about North Korea's efforts to develop multiple-stage missiles. The August 1998 flight test of a *Paektusan*-1, or *Taepo-dong*-1 (TD-1) missile – configured as a three-stage space launch vehicle – was a mixed success. It achieved some technical milestones, but fell short on others. In any event, the TD-1 would not be capable of delivering a nuclear payload to intercontinental ranges. On paper, the larger *Taepo-dong*-2 (TD-2) is a much more capable missile, with a possible strategic capability against the continental US. However, it is not possible to determine how much progress Pyongyang has made on the TD-2 missile or even whether it is ready for testing. Since September 1999, North Korea has observed a moratorium on long-range missile testing, which imposes significant limits on Pyongyang's ability to develop or deploy multiple-stage systems. Given the importance that North Korea has attached to missile development – whether as a strategic asset, bargaining chip, or both – it is plausible that such work continues. However, the true status of North Korea's efforts to develop an intercontinental ballistic missile with a nuclear capability remains unknown.

From FROG to *Scud*-derivatives: a decade of short-range missile development

In 1968, the Soviet Union supplied North Korea with FROG rockets, Pyongyang was probably able to produce FROG copies soon after.[1] The FROG system had strategic utility to Pyongyang because these solid-propellant ballistic artillery missiles, with a range of 50–60km and with a 400kg high-explosive or chemical warhead, can hit Seoul from North Korean firing positions near the Demilitarised Zone (DMZ). This move presaged Pyongyang's efforts, in the mid-1970s, to acquire or develop other short-range ballistic missiles. Meanwhile, education and training programmes designed to generate North Korean expertise in missile-related disciplines had begun before that time.

Pyongyang's efforts to develop systems with a longer range than the FROG, with the intention of striking targets located deep in South Korean territory – presumably including cities as well as military-related facilities – may have been partly spurred by a variety of South Korean efforts, beginning in the early 1970s, to develop surface-to-surface missiles derived from US-supplied *Nike-Hercules* surface-to-air missiles. It is

'North Korea's interest in developing a ballistic missile capability appears to stem from its continuing efforts to establish and maintain robust military forces against South Korea, Japan and US forces in the region'

North Korea's Ballistic Missile Programme

widely assumed that Pyongyang also sought the capability to strike targets in Japan, both to intimidate the Japanese government, so that it would not cooperate with the US in the event of a war on the Korean Peninsula, and to strike US military facilities in Japan. Presumably, North Korea's air force capabilities – weak relative to the US and its allies – gave Pyongyang more incentive to develop a ballistic missile capability to strike distant targets, just as Arab states saw the acquisition of ballistic missiles as a strategic

Ballistic missile ranges	
DESIGNATION	**RANGE**
Short-range Ballistic Missile (SRBM)	Up to 500km
Intermediate-range Ballistic Missile (IRBM)	500–5,000km
Intercontinental Ballistic Missile (ICBM)	Over 5,000km

This table is extracted from the *The Military Balance 1999•2000* (Oxford: Oxford University Press for the IISS)

counterweight to Israel's air superiority after the 1967 Six-Day War.

By the early 1970s, the Soviet Union had begun to export short-range *Scud* ballistic missiles to its main Arab allies in the Middle East, and the missiles played a minor role in the 1973 Yom Kippur War. Strained political relations between Moscow and Pyongyang, however, apparently precluded similar exports to North Korea. Instead, Pyongyang turned to Beijing. During a visit by North Korean leader Kim Il Sung to Beijing in April 1975, North Korea reportedly proposed the development – with China – of a short-range ballistic missile, using the expertise China had gained in developing its medium- and long-range ballistic missiles.[2] The resulting missile project, designated the *Dong-feng*–61 (*East Wind*–61) envisioned a single-stage mobile missile with a range of 600km and a one-tonne (1,000kg) payload. Probably based on Soviet *Scud* technology, which China had obtained from the Soviet Union, the planned missile was designed to use storable liquid fuels, turbo pumps, and inertial guidance. Work on the *Dong-feng*–61 may have started in 1977, but the project was suspended in 1978 when its main supporter General Chen Xilian was ousted from

the Chinese government. Nonetheless, North Korean technical involvement in the programme may have assisted subsequent efforts by Pyongyang to build its own copy of the *Scud* missile.

About the same time that the *Dong-feng*-61 project collapsed, North Korea formed a strategic partnership with Egypt that allowed North Korea to reverse-engineer *Scud* missiles. The relationship was borne out of mutual necessity: the collapse of Egypt's military supply relationship with the Soviet Union following the 1978 Camp David accords had deprived it of the external assistance necessary to maintain and develop its ballistic missile force; Pyongyang had lost a potential ballistic missile source with the collapse of the *Dong-feng*-61 project. Between 1979–1981 – the exact date is unknown – Cairo provided Pyongyang with a small number of Soviet-supplied *Scud*-B missiles, along with mobile launchers, which North Korea could use as a basis for developing an indigenous capability to produce short-range ballistic missiles.

The *Scud*-B missile, a relatively simple and rugged system, was suitable because it both met Pyongyang's military requirements and could be reproduced within North Korea's technological capabilities. Based on German V-2 technology, the *Scud*-B missile is a single-stage liquid-fuelled ballistic missile with a range of 300km and a 1000kg payload – a significant improvement over the range and payload of the solid-propellant FROG artillery rockets in North Korea's existing inventory. The *Scud*-B type provided to Egypt by the Soviet Union is designed to be transported and launched from a mobile launcher and carries a unitary high-explosive warhead of about 800kg, which does not detach from the missile body when the engines stop burning – thereby avoiding the technical complications of warhead separation that are typical of longer-range systems. The *Scud* steel airframe, a little more than 11m in length and 88cm in diameter, contains separate tanks of fuel (kerosene) and oxidizer (nitric acid), which are burned in the missile engine to generate thrust. These basic ingredients are easy to produce. The fact that the liquid fuels used can be stored also simplified design and operation over the use of cryogenic fuels, which had to be maintained at very low temperatures, employed in other types of liquid-fuelled missiles. The *Scud*-B is steered by four flaps, or jet vanes, that protrude into the exhaust plume and divert some of the thrust laterally – a method used by many early generation missiles. With a relatively unsophisticated on-board guidance system, the *Scud*-B has an accuracy of about 1km circular error probable (CEP).[3]

'Based on the missiles and mobile launchers provided by Egypt, Pyongyang embarked on a national programme in the early 1980s to build an indigenous version of the Scud-B missile, thought to be designated the Hwasong-5 (Mars-5)'

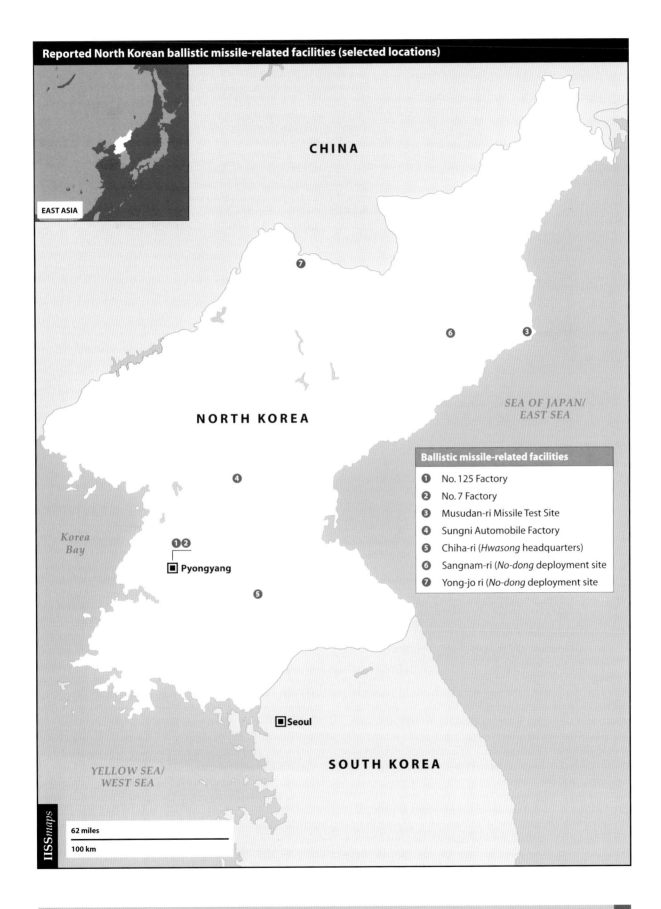

Reported North Korean ballistic missile-related facilities (selected locations)

Ballistic missile-related facilities

1. No. 125 Factory
2. No. 7 Factory
3. Musudan-ri Missile Test Site
4. Sungni Automobile Factory
5. Chiha-ri (*Hwasong* headquarters)
6. Sangnam-ri (*No-dong* deployment site)
7. Yong-jo ri (*No-dong* deployment site)

62 miles

100 km

Reported North Korean ballistic missile-related facilities (selected locations)

1. No. 125 Factory ('Pyongyang Pig Factory')	Reported to be North Korea's primary assembly facility for *Hwasong*-5/-6 and *Nodong* missiles.[6] Various missile components such as engines, airframes, and guidance systems are manufactured at other, unknown locations and transported here for final assembly. Military delegations from several Middle Eastern countries have reportedly toured the assembly plant.
2. No. 7 Factory (Sanum-dong Research and Development Centre)	Reportedly the main ballistic missile research and development facility in North Korea; produces prototypes of missiles and conducts performance tests.[7] Mockups or prototypes of *Taepo-dong*-1/-2 missiles have been detected here on satellite imagery.
3. Musudan-ri Missile Test Site	North Korea's major test launch facility, used to flight test the *Hwasong*-5/-6, *No-dong* and *Taepo-dong*-1 missiles. Based on commercial satellite imagery, the site covers about 9 square kilometres, comprising one launch pad and tower, a missile assembly building, a range control facility and a cluster of four or five small buildings, as well as a facility for static engine tests for missile engines under development.[8]
4. Sungni Automobile Factory	North Korea's main production facility for civilian cars, trucks and buses. Reportedly produces ballistic missile transporter-erector launchers (TELs), mobile-erector launchers (MELs) and support vehicles for *Hwasong*-5/-6 and *No-dong* missiles.[9]
5. Chiha-ri (*Hwasong* headquarters)	Reportedly the headquarters and main technical support base for *Hwasong*-5/-6 forces, built in the late 1980s.[10]
6. Sangnam-ri (*No-dong* deployment site)	Reportedly one location of an underground missile launching site, built in the mid-1990s, presumably for *No-dong* deployment.[11]
7. Yongjo-ri (*No-dong* deployment site)	Reportedly another location for underground *No-dong* deployment, built in the mid-1990s.[12]

NOTE: Defectors and press reports have identified many other sites and facilities reportedly related to missile production and deployment. Much of this information cannot be confirmed. For further information on additional locations, see the Nuclear Threat Initiative, www.nti.org/e_research/profiles/NK/Missile and the Federation of American Scientists, www.fas.org/nuke/guide/dprk/facility/missile

Based on the missiles and mobile launchers provided by Egypt, Pyongyang embarked on a national programme in the early 1980s to build an indigenous version of the *Scud*-B missile, thought to be designated the *Hwasong*-5 (*Mars*-5). Logic dictates that Pyongyang must have built new facilities or converted existing facilities for the production of essential materials and components, including kerosene and nitric acid oxidizer, internal steel tanks and airframes, missile engines, guidance systems and warheads, although much of the publicly available information on production facilities claimed to be associated with ballistic missile production is unsubstantiated.[4] Identifying facilities uniquely associated with the production of ballistic missile components is especially complicated because some components are probably manufactured at facilities primarily responsible for building tactical missiles or that are engaged in some

other military or civilian function. Based on limited defector information and satellite surveillance, key facilities probably include the Sanum-dong research and development facility (also known as No.7 Factory), the Musudan-ri flight test facility, the Sungni Automobile Factory (for mobile launcher production), and the No.125 Factory near Pyongyang (for final missile assembly). North Korean production of the *Hwasong*-5 (and later missiles) was almost certainly supplemented by purchases of specialised materials, equipment, and components from foreign sources.

In 1984, North Korea conducted a series of tests of *Hwasong*-5 prototypes from the Musudan-ri flight-test facility, reportedly totaling three successful and three failed launches.[5] Serial production of the *Hwasong*-5 probably began around 1985–86 and continued until around 1991–92, when serial production of the extended range *Scud*-C (*Hwasong*-6) missile began, most likely using the same facilities, materials and equipment previously used for the *Hwasong*-5. Early versions of the *Hwasong*-5 were delivered to Iran for use during the Iran–Iraq War, which provided the opportunity for North Korean engineers to collect valuable data on the operational use of this missile and to improve production. Based on essentially the same airframe as the *Hwasong*-5, the *Hwasong*-6 was designed to achieve a longer range (of 500km) by reducing the payload from 1,000kg to around 700–800kg, expanding the size of the internal fuel tanks, and slightly modifying the engine for longer burn time.[13] The extended range was of particular interest to Pyongyang, since such missiles could reach all targets in South Korea and also meet the demands of Middle Eastern customers for a longer-range missile. The CEP of the *Hwasong*-6 is unknown, but the missile was probably less accurate than the shorter-range *Hwasong*-5 – perhaps in the order of 1–2km.[14] Pyongyang reportedly conducted five tests of the *Hwasong*-6: in June 1990; July 1991; and May 1993 – the last a multiple test with three missiles. In addition to the *Hwasong*-6, Pyongyang may have worked on other *Scud* variants such as the *Scud*-D with even lighter payloads and longer ranges, but it is not known whether any of these were produced in significant numbers and deployed. Production of the *Hwasong*-6 (and other possible *Scud* variants) may have been phased out in the mid-1990s as North Korea began to build *No-dong* missiles.[15]

There are no reliable figures for annual production capacity and overall production of the *Hwasong*-5 and *Hwasong*-6 missiles in the decade from the mid-1980s to the mid-1990s. No defectors claiming direct knowledge of production statistics have come forward, and intermittent satellite imagery of suspect production facilities cannot determine production rates. Estimates of annual production capacity range from about 50–100 missiles per year, and are based on estimates of North Korean deployments and exports – the presumed total is at least several hundred missiles. North Korea's apparent ability to reverse-engineer sample *Scud* missiles and begin serial production of its own *Scud*-B copy within a few years, and then an extended range *Scud*-C – with a minimal test launch programme – is an impressive technical achievement. This success has led to speculation that the *Hwasong*-5/-6 missiles 'produced' in North Korea in this period were actually assembled from imported components.[16] While it is clear that *Hwasong*-5/-6 production depended heavily on imported production equipment, raw materials, and electronic components, no evidence has come to light

that major subsystems were manufactured abroad. One possible explanation for North Korea's rapid progress is that China, which is believed to have obtained detailed blueprints and technical information on *Scud*-type missiles from the Soviet Union in the 1950s – before the schism between Moscow and Beijing – may have passed this technology to North Korea in the 1970s, thus accelerating North Korea's ability to produce its own version of the *Scud*.[17]

In any event, North Korea probably began deploying *Hwasong*-5 missile units in 1985–86, most likely beginning with small numbers and then establishing larger units as more missiles were produced. Presumably, as *Hwasong*-6 missiles came on-line in the early 1990s, they were deployed in the field in place of some *Hwasong*-5 missiles, but since the *Hwasong*-5 and *Hwasong*-6 missiles are virtually identical in external appearance and use very similar types of mobile launcher, the mix of missile types in North Korea's missile forces cannot be exactly determined.

Since North Korean missile technology is based on Soviet technology, the organisation of North Korea's missile forces probably replicates the structure of Soviet *Scud* forces.[18] In this organisational structure, a North Korean missile regiment would be divided into several launch battalions, with a battalion divided into two or three firing batteries, each consisting of two launchers, reload missiles, and support vehicles. Thus, assuming that North Korea follows typical Soviet doctrine, each battalion would control four to six launchers, with each launcher accompanied by four missiles, one on the launcher itself and three in reserve carried by a supply vehicle. Conservatively, North Korea is estimated to maintain a single *Hwasong*-5/-6 missile regiment of about 30 mobile launchers making for a total of about 120 missiles deployed within the regiment, not counting missiles in reserve.[19]

Although the total number of *Hwasong*-5/-6 missiles (including those stored in reserve) is unknown, they probably total several hundred. In 2000, the US Department of Defense estimated that North Korea's 'ballistic missile inventory now includes over 500 *Scuds* of various types' – an assessment shared by the South Korean Ministry of National Defense. However, this number is based on rough assumptions, rather than direct information.[20] For example, it is possible to count the number of shelters and bunkers assumed to be associated with *Hwasong*-5/-6 missile launchers, but it is impossible to determine how many of these shelters are in use at any one time or indeed how many may be decoys.

'Conservatively, North Korea is estimated to maintain a single Hwasong-5/-6 missile regiment of about 30 mobile launchers making for a total of about 120 missiles deployed within the regiment, not counting missiles in reserve'

North Korea's Ballistic Missile Programme

Defector information on North Korea's missile programme		
NAME	**BACKGROUND**	**INFORMATION PROVIDED**
Lim Ki Sung	Missile scientist trained in Russia; worked in the North Korean missile programme. Defected December 1999.	North Korea has developed a missile with a range of 6,000 km. He was sent to a Chinese missile base in the mid-1990s to observe operations for launching missiles at Taiwan.[23]
Chung Gap Yul	Director of a 'sound research institute', worked on rocket development in the early 1990s. Defected in 1996.	North Korea acquired missile technology by analysing missiles bought from the former Soviet Union. Physicist Suh Sang Wuk is the top supervisor of North Korea's missile industry.[24]
Chang Sung Gil	North Korean Ambassador to Egypt. Defected in August 1997.	Supplied information on North Korea's missile exports to Egypt and the Middle East.[25]
Kim Tae Ho	Worked in a 'chemical factory' under the Ministry of Atomic Energy. Defected in April 1994.	North Korea has built a missile base in Hwadae, North Hamgyong Province, with warheads aimed at Tokyo. Former Soviet missile technicians visited the base.[26]
Bok Koo Lee (pseudonym)	Worked as a missile scientist between 1988 and 1997 at Munitions Plant No. 39 in Huichon, Chagang Province; headed Technical Dept. at Subplant 603. Defected in July 1997.	There are 11 Subplants to Munitions Plant No. 39 of which number 603 and 604 produced and assembled missile parts.[27] Over 90 percent of missile guidance system parts came from Japan, and North Korean missile production dropped about 30% in the aftermath of the *Taepo-dong* launch in 1998, when Japan imposed tighter export controls.[28] He witnessed a missile test conducted in Iran in 1989.
An Yong Chol (pseudonym)	North Korean army general. Defected in 2002.	North Korea has four Soviet-made ICBMs with a range of 8,000 km and dozens of nuclear weapons. Missiles are stored underground at the foot of Mount Paekdu in Potaeri, near the Chinese border.[29]
Kenki Aoyama (pseudonym)	Worked as a missile engineer in North Korea, and then as a 'spy' in China. Defected in 1999.	North Korea acquired nuclear technology in return for providing Pakistan with missile technology.[30] In 1993, Kim Il Sung ordered construction of a missile base between Kungmang Peak and Mujung Peak near Yongjo-ri. Base consists of 12 tunnels, six for missiles and six for personnel and supplies. Each missile tunnel has five or six compartments, resulting in a total storage capacity of some 36 missiles.
Ko Chong Song	Worked as a 'guidance official' at the Kim Il-Song Historical Museum in Chagang Province. Defected June 1993.	The No. 26 Factory is North Korea's largest underground missile facility, built by tunnelling into a mountain. The plant employs 20,000 staff and produces surface-to-air, air-to-surface, and air-to-air missiles, named *Hwasong*-1,-2, and –3.[31]
Ko Young Hwan	Official in Ministry of Foreign Affairs, 1978 to 1991. Defected in 1991.	His older brother was a missile engineer who worked at the January 18th Machinery Factory in Kagam-ri, Kaechon in South Pyongan Province. According to Ko's brother, the underground facility employed 10,000 staff and produced engines for missiles, rocket ships, tanks and torpedoes, including reverse-engineered *Scud* engines since the late 1970s. In 1988, his brother was transferred to the missile engine design lab of the National Defense Institute at Yongseung District in Pyongyang, where he worked on enhanced *Scud* missiles. There is a missile base in North Hamgyung Province, in the Hwadae area, which targeted the US armed forces in Japan[32]

Defector information on North Korea's missile programme (continued)

NAME	BACKGROUND	INFORMATION PROVIDED
Choi Ju Hwal	Former Colonel in the Ministry of the People's Army (KPA) between 1968 and 1995. Defected in 1995.	In 1997, he said that North Korea had developed and deployed missiles such as the *No-dong*-1 and *No-dong*-2 with a range of 1,000km, and was close to finalising *Taepo-dong* missiles with a range of 5,000km. Identified numerous rocket factories in North Korea, including the 125 Factory in the Hyengjesan Area of Pyongyang, the No. 26 Factory in Kanggye of Chagang province, the Yakjeon Machinery Factory in Mangyongdae-ri and the January 18 Factory in Kagam-ri, Kaechon in South Pyongan Province. Identified brigade-sized missile bases in Sangwon in Pyongyang and Hwadae in North Hamgyong Province.
Im Young Sun	Commander of Guard Platoon in the Military Construction Bureau (MCB) of the People's Armed Forces Ministry. Defected in August 1993.	The Military Construction Bureau (MCB) completed the construction of missile bases at Mt. Kanggamchan in South Pyongan Province in 1985; at Paekun-ri in North Pyongan Province in 1986; at Hwadae in North Hamgyong Province in 1988; on Mayang Island in the late 1980s; and at Chunggangjin in Chagang Province in 1995.[33] Some of the bases deployed missiles targeted at Japan and US troops in Okinawa. In 1991, the MCB started to construct an underground missile base at Ok-pyong Rodongja-ku in Kangwon Province and this base was to be completed within 6–7 years. A delay in missile development brought the construction of the Shingye *Scud* Missile Base to a halt, and a long-range missile base is located at Wonsan.
Kim Kil Son	Worked for 20 years for 'highly-classified' Publishing House of the Second Academy of Natural Sciences, North Korea's primary strategic weapons R&D organisation. Defected in 1997.	The Engineering Research Institute under the Second Academy of Natural Science is responsible for the development of ballistic missiles. Most missile-related production facilities are concealed as 'light electric plants' which can be moved from one underground tunnel to another.[34] North Korea's first short range ballistic missile, called the *Hwasong*-1, became operational in 1981.[35] North Korea's key ballistic missiles test facilities are located in Musudan-ri.[36] The No. 7 Factory produces prototypes of missiles and conducts performance tests. Missiles are mass produced at the Mangyongdae Electric Machinery Factory, which is located underground and employs 1,500 staff. Construction of this factory started in 1965 and it was completed in 1978. There are about 200 other factories that produce various missile parts and components.[37]
Ho Chang Kol	Worked in the North Korean 'Speed Battle Youth Shock Brigade'. Defected in November 1996.	The No. 301 Factory in North Pyongan Province in Taegwan County is an underground factory that produces surface-to-air and surface-to-surface missiles and employs some 7,000 staff.[38]
Yi Chung Kuk	Sergeant in the nuclear-chemical defence department under the General Staff of the Ministry of People's Armed Forces. Defected in March 1994.	Missile bases are located in Myongchon and Hwadae of North Hamgyong Province, capable of striking Okinawa and Guam. Missiles at a base in Jagang Province, which borders China, are aimed at China.[39]

NOTE: No effort has been made to verify the claimed backgrounds and information provided by these individuals. The chart is intended to illustrated the range and type of information provided by defectors.

North Korea's Ballistic Missile Programme

Notional organisation of North Korea's *Hwasong*-5/-6 missile regiment (based on Soviet model)

North Korea's operational *Hwasong*-5/-6 missile force is assumed to be organised along the same lines as Soviet *Scud*-B/-C forces since the technology is essentially the same. In one notional configuration, a *Hwasong*-5/-6 regiment could be divided into five launch battalions and each launch battalion could consist of three firing batteries, each with two missile launchers. Thus, each battalion would control six launchers and the regiment would control 30 missile launchers. The total number of missiles deployed in this configuration would be 120, assuming four missiles per launcher.

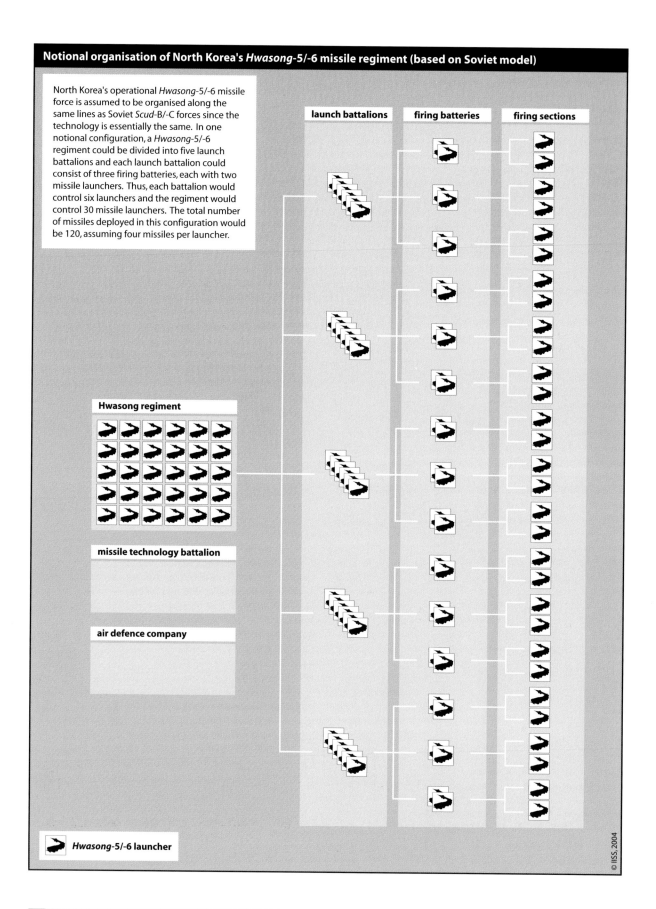

launch battalions

firing batteries

firing sections

Hwasong regiment

missile technology battalion

air defence company

➤ *Hwasong*-5/-6 launcher

© IISS, 2004

The *Hwasong*-5/-6 regimental headquarters is believed to be located in the Chiha-ri area (about 50km north of the DMZ), in a location where satellite imagery has detected a nest of hardened bunkers and tunnels. These are thought to store mobile launchers, missiles and associated equipment.[21] The regiment has reportedly deployed launch battalions and firing batteries in various locations across the country, but, as with the location of missile production facilities, much of the publicly available information from defectors and press stories cannot be confirmed.[22] Washington and Seoul have identified a number of locations, including command centres, bunkers, tunnels and maintenance facilities near the DMZ, that are assumed to be associated with *Hwasong*-5/-6 units. A number of pre-prepared launch sites have also been identified. In wartime, *Hwasong*-5/-6 mobile launchers would be scrambled from hide locations to these sites for firing and then moved as quickly as possible to another location for reloading.

Targeting doctrine and armament for *Hwasong*-5/-6 missiles is uncertain. Some *Hwasong*-5/-6 missiles are equipped with unitary high-explosive warheads and perhaps submunition bomblets, intended for delivery against cities, and military-related command locations, ports, and airports throughout South Korea. Inaccuracy and missile defences would limit the missile's military effectiveness against any target. However, the political impact of, and the civilian terror generated by a number of missiles hitting cities on a daily basis, could create tremendous political pressure on leaders. Along with unitary warheads, North Korea may have been able to develop submunition bomblets to blanket a small target area. Military planners in the US and South Korea assume that chemical and possibly biological warheads are also available, although this cannot be confirmed. North Korea is almost certainly capable of building unitary CBW warheads with various types of agents and impact fusing; whether it has developed more sophisticated means of delivering CBW involving proximity fuses and bomblets is unknown. *Hwasong*-5/-6 missiles are generally thought to have too small a diameter to deliver a first generation nuclear warhead.

Although South Korea is presumed to be the primary target area for North Korea's *Hwasong*-5/-6 missile force, Seoul has typically viewed these missiles as a less serious threat than Pyongyang's more extensive short-range artillery and rocket inventory, which is capable of hitting Seoul with large numbers of high-explosive or chemical rounds. Nonetheless, *Hwasong*-5/-6 hide locations and launchers are likely to

be high priority targets for US and South Korean forces in a conflict. As a consequence, analysts argue that Pyongyang would likely scramble its missile forces as as soon as possible before fighting begins and use its missile forces early in a conflict to avoid pre-emption. It is impossible to accurately estimate how many North Korean missile launchers and crews would survive potential pre-emptive action, but at least some missiles would likely escape such strikes and be launched. Although US and South Korean forces currently deploy *Patriot* Advanced Capability-3 (PAC-3) missile-defence batteries around key military sites in South Korea, at least some missiles that are launched are likely to penetrate these defences. On balance, however, assuming *Hwasong*-5/-6 missiles are not armed with nuclear weapons, the damage they could inflict on major South Korean population centres is probably less than the damage Seoul could suffer from the short-range battlefield and artillery-type systems deployed near the DMZ. The *Hwasong*-5/-6 threat against US military force locations in South Korea is also limited. As a result, Washington and Seoul have not placed the highest priority on North Korean short-range missiles in their negotiating strategies with Pyongyang. However, as US forces along the DMZ are gradually deployed further south, out of North Korean artillery range, Pyongyang may have an incentive to build up its short-range missile forces even further.

Decade of *No-dong* MRBM development

By the early 1990s, the deployment of *Hwasong*-6 missile units gave North Korea the ability to strike targets throughout South Korea with high-explosive or CBW warheads. A new missile design would be needed to reach targets in Japan and to deliver a nuclear warhead. To meet these requirements, North Korea embarked on a programme in the late 1980s to build a new missile, known as the *No-dong*, with a range of 1,000–1,300km and payload of 700–1,000kg.[40] In addition to meeting North Korean security requirements, the missile proved a popular export item, both for existing customers, such as Iran, and for new customers, such as Pakistan.

The *No-dong* missile is essentially a scaled-up *Scud*, using similar or identical liquid propellants with a larger engine. Like the *Scud*, the missile is single stage, but it is roughly 50% larger in length and diameter – with a length of approximately 15–16m and diameter of about 1.2–1.3m. At 16 tonnes, it is also two and half times larger in terms of mass. To achieve its maximum range, the *No-dong* must reach speeds 50% greater than

'A new missile design would be needed to reach targets in Japan and to deliver a nuclear warhead. To meet these requirements, North Korea embarked on a programme in the late 1980s to build a new missile, known as the No-dong'

North Korea's Ballistic Missile Programme

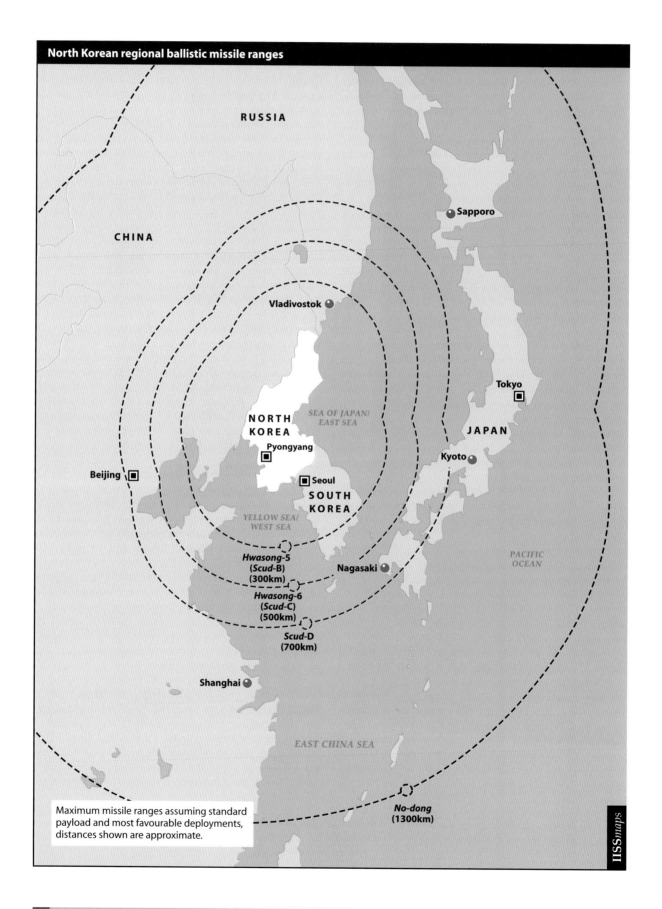

North Korean regional ballistic missile ranges

RUSSIA

CHINA

Sapporo

Vladivostok

NORTH
KOREA
Pyongyang

SEA OF JAPAN/
EAST SEA

Tokyo

JAPAN

Kyoto

Seoul

Beijing

SOUTH
KOREA

YELLOW SEA/
WEST SEA

Hwasong-5
(*Scud*-B)
(300km)

Nagasaki

Hwasong-6
(*Scud*-C)
(500km)

PACIFIC
OCEAN

Scud-D
(700km)

Shanghai

EAST CHINA SEA

Maximum missile ranges assuming standard
payload and most favourable deployments,
distances shown are approximate.

No-dong
(1300km)

IISS*maps*

the *Scud*-C, and doing so with its greater mass requires a new engine that can produce roughly four times the thrust of a *Scud* engine. Interestingly, the design of the *No-dong* is very similar to that of the 'fat *al-Abbas*' or S13 project begun by Iraq in 1989 to develop a missile that could deliver a 1,000kg payload to 1,200km.[41] As far as is known, there was no direct contact between Iraqi and North Korean designers, but both may have been working from old Soviet designs for a scaled-up *Scud*, sometimes referred to as the *Scud*-E project. The concept design was generated by the Makayev design bureau, but it was never developed in the Soviet Union. North Korea may also have had direct assistance from Russian designers in developing the *No-dong* missile, although this cannot be confirmed.[42]

Little is known about the development, production, and deployment of the *No-dong* missile. Development work is assumed to have begun at some point in the mid- to late-1980s. According to press reports, US satellites detected a *No-dong*-type missile at the Musudan-ri test launch facility site in May 1990, and subsequent imagery showed scorch marks on the launch pad, suggesting that the missile had exploded on take off. In May 1993, however, a *No-dong* was successfully tested from the Musudan-ri launch site to a distance of 500km into the Sea of Japan (East Sea). No other *No-dong* tests have been conducted by Pyongyang, although North Korea's missile export agreements with Pakistan and Iran have apparently included provisions for North Korea to participate in tests and to obtain test data from Pakistan and Iran – both of which have conducted a series of missile tests of *No-dong* copies since 1998.[43] There have not been enough *No-dong* tests to provide an accurate assessment of the missile's accuracy: estimates suggest a CEP of 3–5km, but this could be considerably larger if the warhead spirals or tumbles during reentry.[44]

North Korea is believed to have produced *No-dong* units in the same facilities used for the production of *Hwasong*-5/-6 missiles. As with the *Hwasong*-5/6, the line between pilot and serial production of the *No-dong* and the line between partial and full deployment is hazy. Small numbers of *No-dong* missiles may have been produced and deployed in 1994–95, but full production may not have begun until 1996–97, and the system was not considered operationally deployed by South Korea's defence ministry until 1997.[45] Although most of the production capacity is indigenous, North Korea has reportedly obtained some imported materials and components for *No-dong* production such as special steel alloys from China and electronic components from

Japan. The production rate for the *No-dong* missile is unknown, but probably never exceeded an average of about half a dozen missiles per month, with some missiles slated for deployment in North Korea and some for export. Like the *Hwasong*-5/-6, the *No-dong* is designed for delivery from mobile launchers, which North Korea is believed to produce at the Sungni Automobile Factory.

The size, deployment and armament of North Korea's *No-dong* missile force are unknown. Conservatively, *The Military Balance* estimates around ten *No-dong* missile launchers, organised into two or three battalions, but the number could be greater. This would imply at least 40 deployed missiles (assuming that North Korea adopts the Soviet practice of deploying four missiles per launcher in the field), and perhaps a total of 100-200 missiles including reserves.[46] As with estimates of *Hwasong*-5/-6 launchers and missiles, estimates of North Korea's *No-dong* force are very approximate, based on various calculations and assumptions concerning production rates, numbers of shelters and firing points, and military organisation and practice, though the actual number of *No-dong* launchers and missiles cannot be precisely determined. Nevertheless, the size of North Korea's *No-dong* missile force essentially represents Pyongyang's calculation of what it requires in order to meet its military requirements and political needs. In principle, there is no reason why North Korea could not expand its *No-dong* force if it believed it was necessary.

Unlike *Hwasong*-5/-6 missiles, which needed to be deployed near the DMZ in order to reach their presumed targets, *No-dong* missiles can be deployed further north. According to various press reports, satellite intelligence has detected a series of underground launch sites near the Chinese border and the East Sea, which are apparently constructed to allow *No-dong* launchers to enter tunnels and launch through openings at the top of underground bunkers.[47] Given its range, the primary targets for the *No-dong* missile are presumed to be in Japan. Armed with a high-explosive warhead, the *No-dong* is probably not accurate enough for effective use against military targets, such as US military bases in Japan, but it could serve as a terror weapon against Japanese cities, including Tokyo. North Korea is assumed to be capable of arming *No-dong* missiles with CBW warheads, but there is no reliable information on whether it has done so or on what type of warhead and delivery system it may have employed. The potential effectiveness of a chemical or biological warhead against civilian targets

*'The size of North Korea's **No-dong** missile force essentially represents Pyongyang's calculation of what it requires in order to meet its military requirements and political needs'*

varies significantly depending on such factors as the type of agent, the type of delivery mechanism, and the effectiveness of civil defence efforts. As discussed in the chapter dealing with North Korea's nuclear programme, it is possible that North Korea could arm the *No-dong* missile with a nuclear warhead. There is, however, no confirmation that it has done so.

The ability of *No-dong* missiles to successfully attack targets in Japan depends upon the survivability of North Korean missile forces against offensive action, as well as the ability of *No-dong* missiles to penetrate whatever missile defences they encounter. Given that *No-dong* missiles could be armed with nuclear weapons, US and allied forces are likely to make every effort to destroy bunkers and facilities suspected of hiding these missiles and their launch-infrastructure before they can be mobilised and used. On the assumption that suspected *No-dong* bunkers and launch sites would be a high priority target for US and allied forces in wartime North Korea is credited with taking a number of active measures – for example, the construction of decoy shelters and redundant bunkers – to conceal the size and disposition of its missile force.

There may also be undetected *No-dong* hiding locations. It is presumed that, should a conflict occur, North Korea would plan on scrambling its *No-dong* forces before pre-emptive action could be taken against them, and that Pyongyang might fire missiles early in a conflict if the survival of the force were jeopardised.

Nonetheless, given its low numbers and poor accuracy, the *No-dong* is probably more a political weapon than an effective military instrument. In one wartime scenario, for example, North Korea could fire a small number of conventionally armed missiles and threaten to escalate to unconventional warheads if the US and its allies did not accept a cease-fire. Presumably, Pyongyang recognises that the actual use of a nuclear warhead (if it has one) would be suicidal, but it calculates that the threat that it might take desperate measures in extremis serves an important deterrent function.

After the May 1993 *No-dong* test, the US proposed a joint Theater Missile Defense [TMD] development project with Japan, although Tokyo was reluctant to commit itself, in part because of concerns about antagonising China. Following the August 1998 *Taepo-dong*-1 test, however, which aroused a strong public reaction, the Japanese government reached agreement with the US in December 1998 on co-development of a sea-based TMD, called the Navy Theater Wide (NTW)

Pakistani *Ghauri* and launcher (identical to *No-dong*)

defence system. The NTW system envisages a new Standard Missile-3 (SM-3) with a kinetic, hit-to-kill warhead designed to achieve midcourse intercepts outside the atmosphere during an intermediate-range ballistic missile's flight. Given a number of technical issues, the US Navy does not plan initial operational testing of the NTW system until 2010. In December 2003, Tokyo announced that it would proceed with plans to develop an integrated missile defence system, including both sea-based SM-3 and land-based *Patriot* Advanced Capability-3 (PAC-3) missiles.

The *Taepo-dong*-1 shock

By the late 1990s, the *No-dong* missile gave North Korea a credible missile with which to threaten Japan. The *No-dong* aroused particular concern given the possibility that it could carry a nuclear warhead. However, if Pyongyang wanted to pose a threat to US cities with a missile launched from North Korea, or wanted to

'If Pyongyang wanted to pose a threat to US cities with a missile launched from North Korea, or wanted to develop a space launch capability, it would need to develop a missile with considerably greater thrust and range than that of the No-dong'

develop a space launch capability, it would need to develop a missile with considerably greater thrust and range than that of the *No-dong*. Presumably, from Pyongyang's standpoint, a capability to attack US cities might undermine the credibility of US security guarantees to South Korea and Japan in the event of a North Korean attack. Privately, senior North Korean military officers have warned visiting American officials that North Korea is seeking to develop missiles that could attack the US in order to retaliate against any US attack on North Korea. From a technical point of view, the development of the *No-dong*'s engine – more powerful than those used in the *Hwasong* series – was an important step towards developing missiles with even longer range. The other key technology required to attain longer ranges is multiple staging.

The two *Taepo-dong*-family missiles that are assumed to be currently under development both utilise multiple staging. On 31 August 1998, Pyongyang attempted to place a small satellite in orbit by launching a three-stage *Taepo-dong*-1 (TD-1) missile. Although the third stage failed, the test demonstrated for the first time North Korea's technical capability to launch missiles with multiple stages, since the second and third stages both successfully separated. After it seperated, however, the third stage failed to boost the satellite into orbit. By the time it failed, the third stage reached a high enough speed that debris from its breakup reportedly travelled some 4,000km. What caused this third stage failure is unknown. Some analysts have speculated that its engine exploded, while others have suggested that the stage broke apart after tumbling caused by guidance and control failure.

It is not known when North Korea began development work on the TD-1, but in February 1994, US intelligence satellites reportedly detected two new types of missiles or missile mock-ups at the Sanum-dong missile research and development facility.[48] After imagery analysis, estimates were made of the likely staging combinations for the two missiles along with calculations of possible range and payload parameters, based on technical information on the performance characteristics of the *No-dong* and *Hwasong* missiles. US analysts said the press have dubbed these new missile types as the *Taepo-dong*-1 and the *Taepo-dong*-2, although the *Taepo-dong*-1 is apparently known as the *Paektusan*-1 in North Korea.[49] It was originally believed that the TD-1 would be a two-stage ballistic missile, consisting of a modified *No-dong* missile as the first stage and a *Scud*-C as the second stage. Such a combination was estimated to have a range of about 2,000km with a 700–1,000kg

payload, and with an overall mass of 20–25 tonnes. Over the next several years, US analysts sought to follow North Korean efforts to develop these new missile types by, for example, evaluating satellite evidence of static engine tests that appeared to be associated with engines larger than those for the *Scud* missile, but Pyongyang took a number of concealment and deception measures to hide its activities from US intelligence satellites, especially after press leaks of classified information.

Nonetheless, US analysts anticipated, on the basis of estimated development times, that North Korea would be able to test the TD-1 by the late-1990s. As a result, Washington was not surprised when satellites detected preparations for the launch of a TD-1 missile at the Musudan-ri launch site in the summer of 1998, well before the actual launch. Despite warnings from US diplomats not to test the missile, North Korea launched the TD-1 on 31 August 1998. To the surprise of US analysts, the TD-1 included a third stage designed to launch a satellite. The first stage appears to have been a modified *No-dong* missile, which performed in a manner characteristic of a ballistic missile launch. The second stage *Scud* engine, however, appears to have been adjusted for a long burn time at relatively low thrust.[50] Such performance, unlike that of a ballistic missile, is designed to allow the third stage to reach a high altitude before it fires its engine and attempts to put the satellite in orbit.

The third stage consisted of a small solid-fuel rocket motor that carried a small satellite with a mass of probably a few tens of kilograms. The satellite, named *Kwangmyongsong* (*Bright Lodestar*), apparently carried a small transmitter that was intended to broadcast political songs to demonstrate its presence, in a similar manner to the first Chinese satellite in April 1970, which played 'The East is Red'. However, exact information on the origin and technical details of the third stage are unknown. North Korea is not known to have previously developed solid-fuel engines for space launches, but North Korea is capable of building small, solid-fuel rockets for tactical missiles, and it might have obtained assistance from China or other countries in satellite and space launch technology.

The TD-1 launch was perceived as highly provocative, since the launch trajectory was directed to the east of the Korean Peninsula, passing over Japan. However, a launch in that direction is consistent with an attempted space launch, since it allows the missile to use the Earth's rotation to increase its speed. The launch also occurred within days of the 50th

> *'It was originally believed that the TD-1 would be a two-stage ballistic missile, consisting of a modified No-dong missile as the first stage and a Scud-C as the second stage. Such a combination was estimated to have a range of about 2,000km with a 700–1,000kg payload'*

North Korea's Ballistic Missile Programme

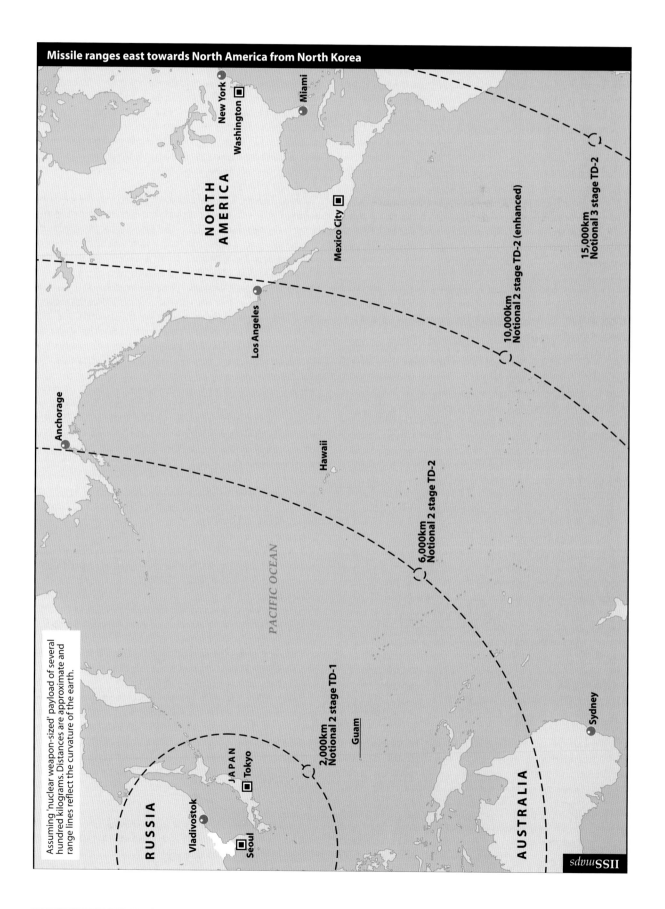

Missile ranges east towards North America from North Korea

Assuming 'nuclear weapon-sized' payload of several hundred kilograms. Distances are approximate and range lines reflect the curvature of the earth.

2,000km
Notional 2 stage TD-1

6,000km
Notional 2 stage TD-2

10,000km
Notional 2 stage TD-2 (enhanced)

15,000km
Notional 3 stage TD-2

Vladivostok

Seoul

Tokyo

JAPAN

RUSSIA

Guam

PACIFIC OCEAN

Hawaii

Anchorage

Los Angeles

NORTH
AMERICA

Mexico City

Miami

Washington

New York

Sydney

AUSTRALIA

IISS*maps*

Selected North Korea missiles – deployed or under development

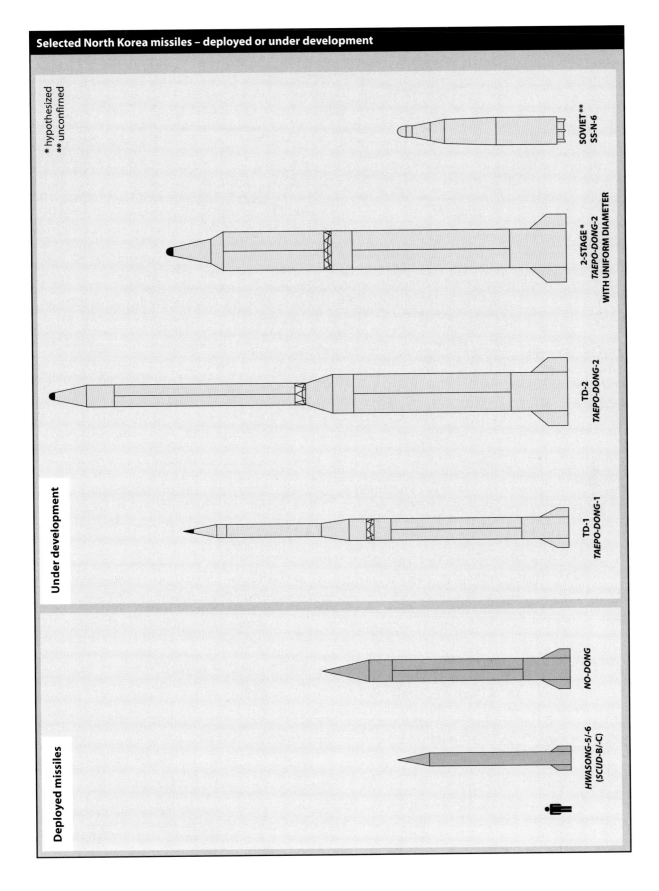

* hypothesized
** unconfirmed

SOVIET **
SS-N-6

2-STAGE *
TAEPO-DONG-2
WITH UNIFORM DIAMETER

TD-2
TAEPO-DONG-2

TD-1
TAEPO-DONG-1

NO-DONG

HWASONG-5/-6
(SCUD-B/-C)

Under development

Deployed missiles

anniversary of the founding of the North Korean state and a consitutional revision that formally transferred power to Kim Jong-Il, suggesting that it was intended to garner prestige and commemorate these events. Citing South Korean and Japanese space launch and satellite programmes, Pyongyang argues that it has the right to develop a space launch capability to orbit its own satellites for telecommunications, meteorological, and other peaceful purposes. In addition, Pyongyang's proclaimed pursuit of a 'peaceful' space launch capability is intended to provide a less provocative political cover for the development of long-range ballistic missiles with a military mission, since the technology involved in space launchers is essentially the same as ballistic missile technology.

In any event, the TD-1 launch caused international uproar, with Japan suspending support for the Agreed Framework and the US threatening to cut off humanitarian food assistance if Pyongyang tested another long-range missile. It also boosted missile defence efforts in the US and Japan, in turn alarming China and Russia, both of whom reportedly warned North Korea against any additional firings. Facing concerted international pressure, Pyongyang embarked on negotiations with the United States, which resulted in agreement in September 1999 on a moratorium on long-range missile tests, in exchange for US agreement to lift a number of economic sanctions. The moratorium is understood to include missiles with, and above, the range of the No-dong. Pyongyang has unilaterally extended the moratorium on several occasions, most recently in September 2002 during the visit of Japanese Prime Minister Koizumi, although North Korea has also threatened to resume 'satellite launches' since it broke the nuclear freeze in late 2002.

The current status of the TD-1 is uncertain. Presumably, Pyongyang has built or could assemble additional TD-1 missiles for firing on fairly short notice, but it is not known whether North Korea has deployed the TD-1 (in a two- or three-stage configurations) to military units. Unlike the No-dong, the TD-1 is not designed for launch by mobile launcher. Some analysts have speculated that it could be deployed in 'silo fashion' in the underground firing bunkers that North Korea has built near the Chinese border. Arguably, however, there would be limited military value in deploying the TD-1 because it contributes little to the strategic role already played by the deployed No-dong force, which effectively covers all critical targets in Japan with a warhead capable of delivering a nuclear weapon. In two-stage configuration, the TD-1 can

deliver a payload comparable to that carried by the No-dong to a greater range, but the extra distance does not encompass any key targets of significant value to North Korea. Moreover, since the second stage of the TD-1 is essentially a Scud missile, the diameter is probably not large enough to accommodate a simple fission-type nuclear warhead, unlike the No-dong.

Using three-stages, the TD-1 is theoretically capable of delivering a small payload of 100–200kg to targets as far away as the continental US, but this payload is considered far less than the amount needed to deliver an early generation nuclear weapon. Even were the missile to carry a chemical or biological warhead, a large fraction of this payload, perhaps two-thirds, would be needed for the structure of the warhead and the reentry heat-shield. In other words, the amount of chemical or biological agent that it could potentially carry would be very small, and probably not enough to pose a significant mass-casualty threat. Moreover, at this maximum range, assessments are that the accuracy of the TD-1 would be very poor. Given these limitations, the TD-1 cannot be considered a serious military threat to the continental US, even if it is deployed. Indeed, it may be the case that the TD-1 was intended primarily as a space launch vehicle and a 'technology demonstrator' to develop North Korean capabilities for the development of longer-range systems capable of striking the US with a militarily significant payload.

Missile defence and the US debate on the Taepo-dong-2
Given the technical limitations of the Taepo-dong-1, the US has long believed that North Korea is much more likely to develop a larger missile, designated the Taepo-dong-2 (TD-2), as a potential intercontinental-range ballistic missile (ICBM) capable of effectively threatening the US with a nuclear-sized payload.[51] While the TD-2 uses components from North Korea's previous missiles, it is significantly different from any missile North Korea has built or tested before and poses a number of new technical challenges. As a result, a number of flight tests would probably be required to develop the TD-2. The TD-2 would use the same engines as the No-dong and the same (or a similar) propellant, but it would require a new, larger first stage, consisting of a cluster of four No-dong engines. This would substantially increase the project's technical complexity and create more opportunities for failure if the four engines do not operate in concert. The TD-2's second stage is thought to use a single No-dong engine, modified for use at high altitudes. Rather than using a No-dong missile body for the second stage,

'Given the technical limitations of the **Taepo-dong-1**, the US has long believed that North Korea is much more likely to develop a larger missile, designated the **Taepo-dong-2 (TD-2)**, as a potential intercontinental-range ballistic missile (ICBM) capable of effectively threatening the US with a nuclear-sized payload'

North Korea is more likely to build a stage that is shorter and greater in diameter than the *No-dong*, since that configuration can reduce the structural mass of the stage, which is crucial for attaining the high speeds needed to reach long ranges.

To achieve maximum range, the TD-2 would need to include a third stage similar to the solid rocket third stage that failed in the August 1998 TD-1 test. North Korea would also need to develop and flight test a re-entry heat shield for a long-range missile before it could use it to deliver a warhead. Overall, the TD-2 missile would be significantly larger than the TD-1, with a maximum diameter (2.4m) nearly twice that of the TD-1 (1.25m). At 75–80 tonnes, it would weigh four times as much as the TD-1 and would generate greater thrust, leading to greater mechanical stresses on the missile's airframe than those seen on previous missiles. As noted above, keeping the structural mass low is crucial to achieving long ranges. However, making the structure sufficiently strong to deal with greater stresses, while minimising overall mass, is a complex engineering challenge.

In the US, assessments of North Korea's TD-2 threat became intertwined with the debate over National Missile Defense (NMD). In general, opponents of missile defence tended to emphasise the technical limitations of the TD-2 and the hurdles that North Korea would face before it could develop and deploy an operational ICBM capable of delivering a nuclear warhead to the continental US. Proponents of missile defence, on the other hand, tended to emphasise the danger of North Korea surprising the US by demonstrating a capability much sooner than expected. In 1995, the public version of US National Intelligence Estimate (NIE) on missile threats concluded that no country, other than the declared nuclear powers, would develop or otherwise acquire a ballistic missile in the next 15 years that could threaten the continental US. Turning to North Korea, the NIE assessed that Pyongyang was unlikely to achieve the technological capability to develop an operational ICBM longer than the 4,000–6,000km-range *Taepo-dong-2*, which could theoretically reach Alaska (but not the rest of the continental US or Hawaii) with a payload large enough to accommodate a nuclear warhead. According to the report, 'For such an ICBM, North Korea would have to develop a new propulsion and improved guidance and control systems, and conduct a flight test program.'[52] Congressional critics of this intelligence estimate believed that it underestimated the potential ballistic missile threat, and Congress passed legislation to form a 'Commission to Assess the Ballistic Missile Threat to the United States'. In July 1998, the Commission concluded that North Korea could develop an ICBM within five years of a decision to do so.[53]

The August 1998 TD-1 space-launch attempt was seen as supporting the Commission's argument that Pyongyang could make significant progress towards the development of a missile that could threaten the US with little or no warning, especially given the inherent difficulties of collecting information on North Korea. Even though the third stage of the TD-1 was unsuccessful, the failure to anticipate or detect North Korean work on third-stage technology led the US intelligence community (already under criticism for underestimating the missile threat) to incorporate 'worst case' scenarios in their assessments of North Korean missile developments. The intelligence community adopted a looser definition of operational deployment, recognising that Pyongyang might be willing to 'deploy' a new missile system even before a full series of flight tests. In addition, US assessments of the possible range and payload performance of the presumed TD-2 have postulated greater capabilities over time taking into consideration possible techniques for enhancing performance. The original 1995 US government assessments of the TD-2 estimated that it had a range of 4,000–6,000km, sufficient to reach Alaska, but not the main Hawaiian Islands, nor the rest of continental United States.[54] The 1999 NIE on *Foreign Missile Developments and the Ballistic Missile Threat Through 2015* – the first official assessment after the August 1998 launch – concluded that a two-stage TD-2 could carry a 'nuclear weapon sized' payload, defined as 'several-hundred kilograms' to Alaska and Hawaii (7,500 km).[55] In 2001, the official US estimate concluded that 'The Taepo Dong-2 in a two-stage ballistic missile configuration could deliver a several-hundred-kg warhead up to 10,000km' – sufficient to strike some coastal areas in the western continental US.[56] Assuming that the TD-2 is designed to carry a third stage similar to that used in the 1998 TD-1 launch, the 1999 NIE estimated that a three-stage TD-2 'could deliver a several-hundred kilogram payload anywhere in the United States'. In 2001, the NIE specified the range at 15,000km, with a payload of several hundred kilograms, which would be sufficient to allow it to reach all of North America.

The greater capabilities attributed to the TD-2 by more recent US estimates reflect a combination of several possible measures that North Korea could take to enhance TD-2 performance in comparison to that

'US assessments of the possible range and payload performance of the presumed TD-2 have postulated greater capabilities over time taking into consideration possible techniques for enhancing performance'

observed during 1998's TD-1 launch. Manufacturing the missile body from significantly lighter materials, such as aluminum magnesium rather than steel – or improving engine performance – are two possible measures. It is not known whether North Korea is pursuing these routes to enhance TD-2 range, but recent US government estimates are intended to cover such potential 'worst case' contingencies. If it used known North Korean capabilities, the TD-2 could have a CEP of tens of kilometers at long range, which could limit its effectiveness against anything except large targets such as cities.

Though it is presumed that North Korean scientists and technicians are working on improved guidance and control systems, there is little information on such efforts. Improving the accuracy of a ballistic missile is not just a matter of improving its guidance system, which is used to aim the missile during the first few minutes of flight, when it is under power. Once the engines burn out, the warhead falls through space, and a significant part of the missile's CEP results from intense atmospheric buffeting as the warhead reenters the atmosphere in the last few minutes before impact. For example, if ablation does not take place symmetrically over the heat shield that is used to protect the warhead from re-entry heating, uneven wear may result in lateral forces pulling the warhead off-course. Moreover, the difficulties in achieving a low CEP increase as the range of the missile increases. For instance, errors in aiming the missile through its initial burn phase leads to greater errors in hitting the target as target range increases. And, since atmospheric forces increase rapidly with speed, the higher speed of longer-range missiles leads to considerably greater re-entry errors. From Pyongyang's standpoint, however, high accuracy would not be required for the TD-2 to perform its presumed strategic function – that of threatening US cities with nuclear attack.

While range, payload, and warhead type are typically discussed as the primary characteristics of a ballistic missile, another key consideration is whether or not the missile includes countermeasures designed to defeat missile defences. If Pyongyang is indeed determined to pursue long-range missiles, it has both aerospace skills as well as a motivation for developing mechanisms intended to defeat missile defences, such as releasing balloon decoys while disguising the warhead by enclosing it in a similar balloon.[57] It is not known whether North Korea is seeking to develop such mechanisms.

The TD-2 has never been flight-tested and its development status is unknown. Since 1998, the US

government has estimated that the TD-2 'may' be ready for flight testing and could be deployed in a 'few years', but this assessment is based on a combination of assumptions about North Korea's missile development capabilities and fragmentary evidence. For example, in 1999, North Korea reportedly modified the Musudan-ri site (from where the TD-1 was launched) by extending the launch gantry so that it could accommodate the taller TD-2 missile, suggesting that the TD-2 was ready for flight testing, or at least that Pyongyang wanted Washington to believe this to be the case.[58] According to press reports, intelligence satellites have also detected several static engine tests since 1999 that might have been associated with development of a TD-2 engine, although it is difficult to determine the type of engine being tested or to evaluate the results of the tests from the information available.[59] In contrast to US government assessments, the Russian government has tended to regard the TD-2 as a 'paper missile', emphasising that North Korea faces significant technical hurdles in developing a missile that could threaten the US. However, these Russian assessments may be influenced by a desire to undercut the justification for missile defence.[60]

In reality, it is impossible to make a confident judgement about the status of the TD-2 because too little information is available. Even North Korea cannot be certain about the capabilities of the TD-2 until it is flight tested, and since September 1999, Pyongyang has continued to observe a moratorium on long-range missile flight tests, though it is unknown whether this is for technical or political reasons. Since the collapse of the Agreed Framework in late 2002, Pyongyang has periodically issued threats to resume satellite launches – for example after Japan launched a reconnaissance satellite in March 2003 – but no preparations for another missile test have been reported at the Musundan-ri launch site. Even if the TD-2 is successfully tested, some experts argue that a series of flight tests would be required to provide a meaningful estimate of reliability and therefore to consider the missile operational in the normal sense. Other experts cite the *No-dong* programme and argue that North Korea might 'deploy' the TD-2 based on one or two successful tests, calculating that the mere possibility of the missile operating would serve a deterrent role.

New developments?

In September 2003, according to press reports, intelligence satellites observed several copies of a new type of intermediate-range missile (or missile mock-up)

'It is impossible to make a confident judgement about the status of the TD-2 because too little information is available. Even North Korea cannot be certain about the capabilities of the TD-2 until it is flight tested'

and mobile launcher near Pyongyang.[61] Presumably based on its observed dimensions and external appearance, the new missile was said to resemble the SS-N-6, an old Soviet-era single-stage, liquid fuelled submarine-launched ballistic missile developed in the 1960s and designed to deliver a payload of about 700kg to a range of 2,400–3,000 km.[62] The SS-N-6 is shorter and fatter than the *No-dong*, with a length slightly under 10m and diameter of 1.65m, compared to the 15–16m length and 1.2–1.3m diameter of the *No-dong*. Of course, it cannot be determined from satellite images whether the new North Korean missile is actually derived from the SS-N-6 or even whether it is real. Some analysts speculated that the missiles were mock-ups intended to be displayed in a military parade marking the 55th anniversary of the foundation of North Korea, but the missiles did not make a public appearance. If it is a new missile project, the estimated range would not be great enough to strike the US. However, the larger diameter would make it easier to deliver an early generation nuclear device, when compared to the *No-dong*.

In conclusion, since the early 1990s, North Korea has actively pursued an interest in developing space-launch vehicles and long-range ballistic missiles, which requires the development of technology beyond that used in the *No-dong*. Most notably, the August 1998 launch of a three-stage TD-1 demonstrated North Korean achievement of some important technical benchmarks necessary for development of intercontinental missiles, such as stage separation, but failure to achieve others, such as a successful third stage powered by a solid-fuel motor. Deployment of a TD-1 would serve little greater military utility than that served by North Korea's existing *No-dong* missile force. However, Pyongyang could presumably resume TD-1 testing on fairly short notice, whether to renew efforts to orbit a satellite, or develop ICBM technology, or make a political statement – or indeed all three.

In contrast to the TD-1, the *Taepo-dong*-2 is a more credible candidate for an intercontinental missile, although it poses a number of additional technical hurdles. In either a two-stage or three-stage configuration, the TD-2 could theoretically deliver a nuclear weapon-sized payload to cities in the United States. There is some evidence that North Korea has continued to work on the TD-2 or other types of long-range missiles since the August 1998 TD-1 test, but information on the level of effort and the level of success is very fragmentary and elusive. As a result, a firm judgement on the status of the TD-2 cannot be rendered because too little information is available.

North Korean missile exports

North Korea has become the world's most prolific exporter of ballistic missiles and related equipment, materials and technology – especially as other potential suppliers, such as China, have gradually withdrawn from the market. Over the past two decades, North Korea has sold at least several hundred *Hwasong*-5/-6 or *No-dong* missiles, as well as materials, equipment, components and production technology, mainly to countries in the Middle East, such as Egypt, Iran, Libya, Pakistan, Syria, the United Arab Emirates (UAE) and Yemen.[63] During that time, North Korea's missile export business has probably earned several hundred million dollars – a significant portion of North Korea's hard currency earnings. Missile deals have also probably included barter arrangements for oil (with Iran) and nuclear technology (with Pakistan), and have provided North Korea with opportunities to test missiles off-shore. In recent years, however, revenues from missile sales may have fallen off. Some of North Korea's longstanding customers, such as Iran, have come close to achieving an independent production capability, reducing their need for North Korean imports and even presenting competition to North Korean sales. Other customers, such as Pakistan, Yemen, Egypt and the UAE and most recently Libya, have come under political pressure from Washington to sever their missile relationship with Pyongyang. Nonetheless, North Korea remains the world's leading exporter of ballistic missiles and related technology.

As part of the original missile partnership established between Pyongyang and Cairo more than 20 years ago, North Korea probably sold spare parts to help maintain Egypt's *Scud*-B force throughout the 1980s. Cairo, however, focused its missile development efforts on a joint project with Argentina and Iraq to develop the solid propellant *Condor* missile, a project which started in 1985. US pressure on Argentina and European prosecutions of companies involved in the project brought an end to the *Condor* project in 1988–89. After this, Egypt apparently turned back to North Korea, which sold production technology, raw materials and key components for extended range *Scud*-C missiles throughout the 1990s. Under pressure from Washington, following a 1996 incident in which Swiss authorities intercepted a large shipment of Egypt-bound North Korean *Scud*-C missile components at Zurich airport, Cairo made commitments to limit missile cooperation with Pyongyang. In 2000, Egypt reportedly approached North Korea in a bid to acquire engines and other components for *No-dong* missiles,

'North Korea has become the world's most prolific exporter of ballistic missiles and related equipment, materials and technology' ... 'In recent years, however, revenues from missile sales may have fallen off'

though US diplomatic intervention with Cairo apparently blocked the complete transfer.

North Korea's missile exports to Iran began during the Iran–Iraq War when North Korea shipped *Scud*-B missiles (designated the *Shahab*-1 by Iran) and mobile launchers to Iran in 1987 for use against Iraq in the 'War of the Cities'. In the early 1990s, North Korea provided Iran *Scud*-C missiles (designated the *Shahab*-2) and helped Iran to establish an indigenous missile production infrastructure, in exchange for money and oil. In 1993, North Korea negotiated with Iran for the sale of *No-dong* missiles, but the exports were delayed, perhaps because of warnings from Washington that *No-dong* transfers to Iran could derail negotiations for the Agreed Framework, which were taking place at the same time. In 1995 however, after the conclusion of the Agreed Framework, North Korea began exporting *No-dong* missiles to Iran and helped Iran develop its own version of the *No-dong*, which Tehran designated the *Shahab*-3. The North Korean sale of *No-dong* missiles to Iran prompted Washington to begin a series of negotiations with Pyongyang, seeking an agreement to end North Korean missile exports, but the two sides were never able to reach agreement on the amount and type of 'compensation' that North Korea would receive for ending missile-related exports. North Korea may have provided *Taepo-dong* technology to Iran, but this cannot be confirmed.

Reports of Syrian negotiations with North Korea for missile sales surfaced in 1989, apparently after China, under pressure from the United States, withdrew from arrangements to help Damascus upgrade its ageing force of Soviet-supplied *Scud*-B missiles. In the early 1990s, North Korea shipped *Scud*-C missiles and production equipment to Syria, but there have been few reports of substantial North Korean sales to Syria since then, perhaps in part because Syria and Iran developed a close missile cooperation relationship during the mid-1990s. North Korean missile sales to Libya came to light in June 1999, when Indian customs officials seized the North Korean ship *Ku Wol San* after they suspected the ship was delivering missiles to Pakistan. In fact, the *Ku Wol San* was destined for Libya, carrying raw materials, test equipment, machine tools, missile components, and blueprints for *Scud* missiles. Since 2000, there have been several unconfirmed reports that North Korea has sold *No-dong* missiles or components to Libya. Tripoli's December 2003 announcement that it will renounce its weapons of mass destruction programmes and ballistic missiles of over 300km range presumably limits future North Korean sales to Libya.

During the 1990s, North Korea also made minor sales of *Scud* missiles to the United Arab Emirates (UAE) and Yemen. These exports attracted US attention, and Washington negotiated agreements with both the UAE and Yemen to end further missile purchases from North Korea. In December 2002, however, Spanish and US naval vessels intercepted the North Korean ship *Sosan* off the east coast of Africa – the ship was found to be carrying 15 *Scud* missiles to Yemen. The ship's cargo was also carrying conventional warheads and 85 drums of 'inhibited red fuming nitric acid', an oxidizer for *Scud* missile fuel. The ship was stopped ostensibly for being 'unflagged', but was later released when Yemen protested to Washington. However, Yemeni President Ali Abdullah Saleh said that the missiles would be used only for national defence, and that it would be the last shipment. Based on evidence uncovered since the US invasion of Iraq, it also appears that Iraq began negotiations with North Korea in 1999 to purchase *No-dong* missiles and production technology, although the deal fell through in 2002 when North Korea decided it was too risky to begin shipments to Iraq via Syria. Baghdad's downpayment was, though, retained by Pyongyang.

Strategically, North Korea's most significant missile customer has been Pakistan, which reportedly agreed around 1997 to trade uranium enrichment technology to North Korea in exchange for *No-dong* missiles and production technology. According to various reports, the missile components and technicians were transported by aircraft from Pyongyang to Islamabad in late-1997, and in April 1998, Pakistan flight tested its version of the *No-dong*, which it designated the *Ghauri*.[64] North Korean technicians apparently continued to work with Pakistan on development of the *Ghauri*, which was further tested in April 1999 and May 2002. Following the US accusation in October 2002 that North Korea was pursuing a clandestine enrichment programme, Washington has increased pressure on Pakistan to end its missile cooperation with North Korea.

Conclusion

Since the mid-1970s, North Korea has pursued the development of ballistic missiles with increasing range, which it has deployed with its armed forces. By the mid-1980s, North Korea had deployed short-range *Hwasong*–5/-6 missiles capable of reaching targets throughout South Korea. By the mid-1990s, it had deployed *No-dong* missiles capable of reaching all of Japan. The size and disposition of North Korea's *Hwasong* and *No-dong* missile forces are uncertain, but

'Tripoli's December 2003 announcement that it will renounce its weapons of mass destruction programmes and ballistic missiles of over 300km range presumably limits future North Korean sales to Libya'

probably includes a few hundred deployed missiles, with additional missiles in reserve. If deemed necessary, North Korea can expand the size of its deployed missile forces. Pyongyang probably views these forces as both a military and political asset. Militarily, the missiles can serve the function of long-range artillery, seeking to disrupt enemy communications and logistics in rear areas and interdicting reinforcements. To some degree, the military effectiveness of North Korea's missile force would be reduced by poor accuracy, limited survivability, and missile defences, but they could make a significant contribution to overall military operations, especially in the early stages of a conflict. As a political tool, North Korea's missiles give it more ability to threaten cities in South Korea and Japan with conventional or unconventional warheads. Whether unconventional warheads have been deployed is unknown, but the possibility contributes to deterrence and intimidation. In addition to their perceived political and military utility for North Korean defence, the sale of missiles and missile technology has been an important incentive for North Korean missile development and production.

North Korea does not have operational missiles capable of striking the US with nuclear weapons, and will likely not be able to develop them as long as it continues its current moratorium on flight-testing. However, the history of Pyongyang's missile development programme suggests that given the time and resources, it will be able to develop missiles with increasing range if it decides to do so. How long successfully demonstrating such missiles might take is unclear, especially since the level and quality of foreign assistance, and the status of current missile development is not known. As a result, widely divergent scenarios for the status of *Taepo-dong*-2 development can be constructed, and assessments of the TD-2 became intertwined with the politically charged debate over US missile defences. In one view, North Korea is seen as working hard to field a missile that can threaten the United States, which Pyongyang views as essential to undermine the US security relationship with South Korea and Japan and to deter US attacks on North Korea. In an opposing view, the TD-2 is designed more for bargaining leverage and trading for political and economic benefits than for military use. In a sense, both are probably true – by developing greater missile capabilities, North Korea can drive up the price for agreeing to restrain or abandon parts of its missile programme and at the same time be in a stronger position to test and deploy such systems if negotiations fail.

'To some degree, the military effectiveness of North Korea's missile force would be reduced by poor accuracy, limited survivability, and tactical missile defences' ... 'As a political tool, North Korea's missiles give it more ability to threaten cities in South Korea and Japan with conventional or unconventional warheads. Whether unconventional warheads have been deployed is unknown, but the possibility contributes to deterrence and intimidation'

North Korea's Ballistic Missile Programme

The Conventional Military Balance on the Korean Peninsula

Overview

The modern political history of the Korean Peninsula is shaped by the legacy of the 1950–53 Korean War. This bitter and costly conflict ended in military stalemate, with North and South Korea continuing to be divided by the 38th parallel. Subsequently, North Korea amassed large and formidable conventional military forces, which are mainly forward-deployed near the Demilitarised Zone (DMZ), seemingly in position to launch an invasion of South Korea. Over the past two decades, due largely to economic decline and lack of financial resources, as well as force improvements in South Korea and the US, North Korea's conventional forces have become relatively weaker, when compared with those of South Korea and the US. As a result, any North Korean option to invade South Korea has become less credible. While causing tremendous damage, a North Korean attack on South Korea would most likely be defeated by a US–South Korean counterattack. Nonetheless, the credibility of North Korea's conventional military forces remains largely intact in terms of their potential to defend the state and to inflict damage on South Korea – especially Seoul – which remains hostage to North Korea's artillery massed along the DMZ.

By the same token, options for US and allied forces to launch pre-emptive strikes against selected military targets in North Korea are fraught with steep risks, even more for any plan for an invasion to overthrow the North Korean regime. The US could probably destroy known nuclear and missile facilities in a pre-emptive strike, but not hidden facilities and weapons that would survive such a pre-emptive attack. In any event, Pyongyang would regard an attack on its strategic assets as a dire threat to its vital interests, and could retaliate in ways that could quickly escalate to a wider conflict. The US and South Korea would likely prevail in a fullscale war, but the human and material costs would be very high – even if unconventional weapons were not employed. In essence, the military standoff that marked the end of the Korean War prevails 50 years on.

Military geography and the disposition of forces

Conceptually, US defence reviews conducted in the 1990s included the Korean Peninsula along with military contingencies in Southwest Asia. Geographically, however, the situation on the Peninsula bears more similarity to that surrounding the former border between East and West Germany, and to Bosnia-Herzegovina, than it does to Kuwait, Saudi Arabia, or southern Iraq. In particular, the dispositions of North Korean and allied (US and South Korean) armed forces deployed on the Peninsula are dictated by terrain. In geographical terms, the Korean Peninsula is compact, approximately 250km wide at its narrowest point and about 1,000km long. Moreover, the Peninsula is characterised by mountainous topography; much of the flat land that exists comprises either marshland or rice fields. Therefore, rapid movement by heavy armoured forces would be difficult.

The DMZ is approximately 4km wide and 250km long, stretching from the Yellow Sea (or West Sea) in the west to the Sea of Japan (or East Sea) in the east. Contrary to its name, this zone is located within one of the world's most heavily militarised areas. More than one million troops and 20,000 armoured vehicles and artillery pieces – plus more than one million landmines and numerous fortified defensive positions – are packed into a small area surrounding the DMZ. Furthermore, there is little 'strategic depth' between the DMZ and the capital cities of Pyongyang (about 125km north of the DMZ) and Seoul (approximately 40km south of the DMZ). As a comparison, forces on either side of the DMZ are more densely concentrated than were those of the Warsaw Pact and the North Atlantic Treaty Organisation (NATO) in Central Europe during the Cold War.

The region's topography offers heavy armoured forces three main avenues of approach for any potential land offensive. Two are in the relatively flat western part of the Peninsula, known as the Chorwon and Kaesong Munsan corridors, and provide the most direct approaches to Seoul and Pyongyang, although much of the flat terrain is marsh land and rice fields. The third route runs along the east coast through the Taedong Mountains and is the most amenable to vehicle passage. In some places, these corridors are about 15km wide and interconnected with other possible routes, which would utilise existing road networks and suitable terrain in the central and eastern parts of the Peninsula.

North Korean military capabilities

Although it is difficult to know North Korea's precise intentions or aspirations, its forces are deployed along the DMZ in such a manner that they could support an invasion of South Korea. In particular, the percentage of North Korean forces deployed within 100km of the DMZ has significantly increased during the past two decades. Currently, North Korea deploys approximately 65% of its military units, and up to 80% of its estimated aggregate firepower, within 100km of the DMZ. This inventory includes approximately 700,000 troops, 8,000

'The percentage of North Korean forces deployed within 100km of the DMZ has significantly increased during the past two decades'. 'Its forces are deployed along the DMZ in such a manner that they could support an invasion of South Korea'

The Conventional Military Balance on the Korean Peninsula

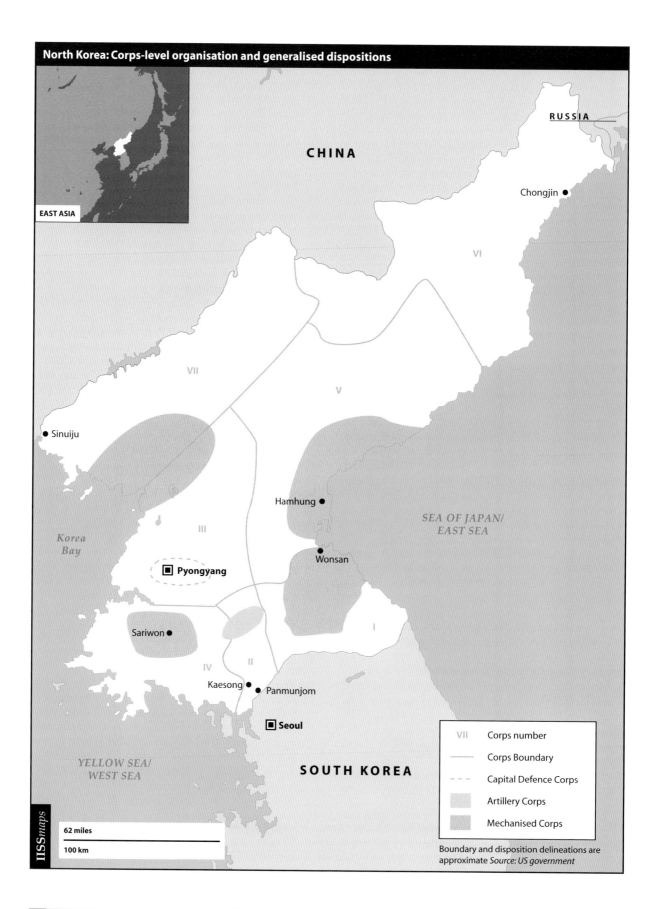

North Korea: Corps-level organisation and generalised dispositions

EAST ASIA

RUSSIA

CHINA

Chongjin ●

VI

VII

● Sinuiju

V

Hamhung ●

Korea
Bay

III

SEA OF JAPAN/
EAST SEA

☐ Pyongyang

● Wonsan

Sariwon ●

I

IV

II

Kaesong ●
● Panmunjom

☐ Seoul

VII	Corps number
———	Corps Boundary
- - -	Capital Defence Corps
▨	Artillery Corps
▨	Mechanised Corps

YELLOW SEA/
WEST SEA

SOUTH KOREA

IISS*maps*

62 miles

100 km

Boundary and disposition delineations are
approximate *Source: US government*

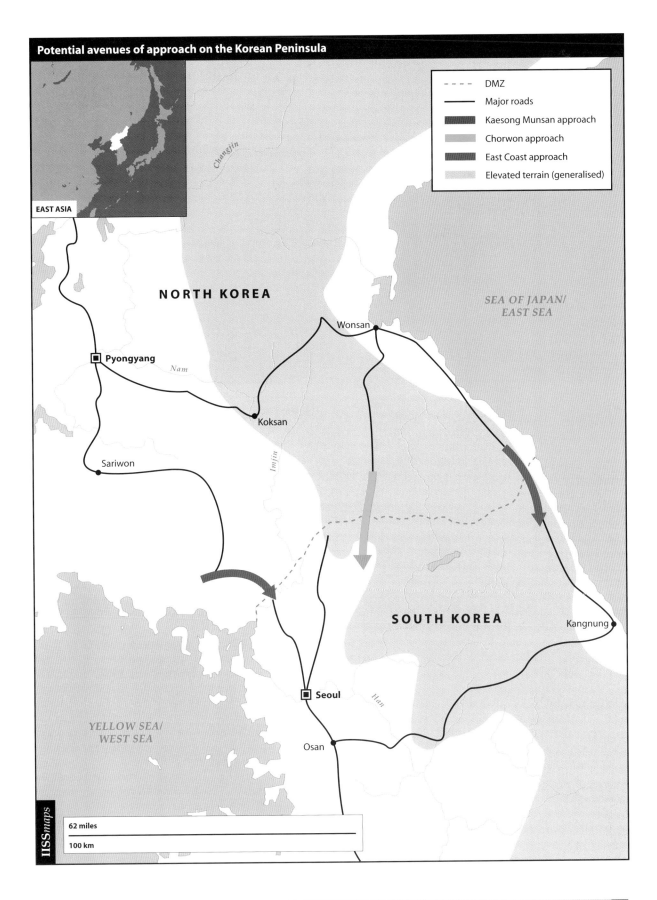

Potential avenues of approach on the Korean Peninsula

EAST ASIA

Legend:
- - - - DMZ
— Major roads
Kaesong Munsan approach
Chorwon approach
East Coast approach
Elevated terrain (generalised)

NORTH KOREA

Changjin

SEA OF JAPAN/
EAST SEA

Wonsan

Pyongyang

Nam

Koksan

Sariwon

Imjin

SOUTH KOREA

Kangnung

Seoul

Han

YELLOW SEA/
WEST SEA

Osan

62 miles

100 km

IISS*maps*

North Korean 70-tonne *Yugo*-class submarine, captured by South Korean naval forces, June 1998

artillery systems and 2,000 tanks. Because of these forward deployments, North Korea could theoretically invade the South without recourse to further deployments and with relatively little warning time.

Thus, it has been argued that North Korea's military strategy is designed around plans to launch an invasion of South Korea. At the same time, North Korea's armed forces are also positioned in order to deter an attack, being deployed to deliver a pre-emptive strike against the South if Pyongyang believes that an attack is imminent or to retaliate with overwhelming force if the North is attacked. This posture is dictated by the doctrine that 'attack is the best form of defence', a formulation that defined Soviet forward deployments in East Germany during the Cold War. The mass forward deployment of North Korean forces also helps to strengthen domestic political support for Pyongyang's 'military first' policy and heavy internal security apparatus.

North Korea, having adopted generic Soviet military doctrine, would probably begin any war with a massive artillery assault on South Korean and US positions south of the DMZ and on Seoul itself.[1] Chemical weapons might also be employed against military and civilian targets. Infantry and mechanised forces would then try to take advantage of the ensuing chaos to penetrate South Korean–US defences with the aim of quickly capturing Seoul. These advancing forces could be preceded by Special Forces, pre-deployed in South Korea through tunnels or inserted by mini-submarines or airdrops. Presumably, North Korea

would try to catch US and South Korean forces off-guard, attempting to seize Seoul before reinforcements could be deployed. Once Seoul was captured, North Korea might try to invade the rest of the Peninsula or try to use Seoul as a bargaining chip to negotiate favourable ceasefire terms. In any event, mass artillery and armoured forces, supported by Special Forces and airborne missions, would be the central components of any offensive operation.

Orbats and equipment

North Korea's armed forces are composed of nearly 1.1m active-duty personnel and some 4.7m reserves, making them the world's fifth largest active military force. Although precise conversion rates for the North Korean Won to the US dollar are difficult to ascertain, North Korea officially maintains an annual defence budget of about $1.5bn to support these forces, but some estimates of actual expenditure are more than three times as high, at around $5bn, which would translate to about 25% of North Korea's GDP, estimated to be currently $20bn.

Pyongyang's order of battle is equivalent to approximately 150 active duty brigades. That includes 27 infantry divisions, as well as some 15 independent armoured brigades, 14 infantry brigades, and 21 artillery brigades.[2] North Korean forces are heavily dug-in with more than 4,000 underground facilities and bunkers near the DMZ and an estimated 20 tunnels dug under the DMZ, of which four have been found. There are also more than 20 Special Forces brigades, totaling about 88,000 troops, which could be deployed by air, sea and land to disrupt US and South Korean combat operations and attack civilian targets.

North Korea's armoured forces are estimated to include some 3,500 main battle tanks (MBTs), 3,000 armoured personnel carriers and light tanks, and more than 10,000 heavy-calibre artillery pieces, many of which are self-propelled. The MBT force primarily comprises older T-54/55/59 models, but includes some 800 indigenously produced T-62s. Of the estimated 10,000 or so artillery pieces in the North Korean inventory, a considerable number are pre-deployed, in range of Seoul; additional artillery could be moved forward to fortified firing positions at short notice. Of particular concern to Seoul are Pyongyang's 240mm multiple rocket launchers (capable of simultaneously firing 16–18 rockets), its 152mm and 170mm towed and self-propelled artillery pieces, and its mobile FROG systems – all of which are capable of delivering chemical and biological agents as well as conventional high-explosives. In addition, the ground forces have about 7,500 mortars, several hundred surface-to-surface missiles, 11,000 air defence guns, 10,000 surface-to-air missiles, and numerous anti-tank guided weapons.

The North Korean air force possess some 605 combat aircraft and is organised into 33 regiments: 11 fighter/ground attack; two bomber; seven helicopter; seven transport; and six training regiments. The air force mostly comprises older MiG aircraft (of the MiG-15/17/19/21 types), but includes small numbers of more modern MiG-23, MiG-29 and Su-25 aircraft. Like North Korea's ground forces, a relatively large percentage of the air force is deployed near the DMZ – at military air bases only minutes flying time from Seoul. The North Korean navy can be divided into six main groups: 43 missile craft; about 100 torpedo craft; 158 patrol craft (of which 133 are inshore vessels); about 26 diesel submarines of Soviet design; 10 amphibious ships; and 23 mine countermeasures ships. There are also some 65 miniature submarines for the insertion and extraction of Special Forces. Around 60 percent of the North Korean navy is deployed in forward bases, and North Korea has strengthened its coastal defences in forward areas by deploying more modern anti-ship cruise missiles.

On paper, North Korea's armed forces are formidable, but their actual capabilities are less than the raw data suggest, given the obsolescence of most North Korean equipment. Around one-half of North Korea's major weapons were designed in the 1960s; the other half are even older. Also, it is certain that due to shortages of spare parts, fuel, and poor maintenance, some weaponry will not be functional. The US Army's Cold War system for comparing hardware capabilities suggests that ground combat units equipped with modern Western weaponry are about 20–40% more combat effective than units of comparable size with out-of-date equipment. After the 1991 Gulf War, the US-based Analytic Sciences Corporation developed a more up-to-date and realistic model for comparing forces, known as the Technique for Assessing Comparative Force Modernization (TASCFORM), which was utilised in the 1990s by the Pentagon's Office of Net Assessment. According to this model, modern Western weaponry is generally two-to-four times more capable than Soviet systems.[3]

Using TASCFORM methodology, it is estimated that North Korea's heavy armoured forces, possessing enough combat hardware to equip perhaps ten US divisions, have an actual capability equivalent to about 2.5 US armoured divisions. With equipment operated by the infantry added, the North Korean ground forces possess an overall firepower which is equivalent to nearly five modern US heavy divisions. By comparison, Iraq was assessed as having six modern division equivalents when the same TASCFORM scoring system was used in 1990. Using the same methodology, North Korean airpower, the equivalent to six US wing equivalents in size, corresponds to only two F-16 wing equivalents in estimated net capability.

North Korean doctrine, military readiness and morale are also key factors in determining actual military performance. Employing highly inflexible Soviet-style military doctrine, North Korea emphasises high-ranking decision-makers and scripted war plans – neither of which encourage operational flexibility nor initiative. It is doubtful, therefore, that North Korea possesses a strong mid-level officer corps. Pyongyang has attempted to raise training levels and readiness in recent years, but fuel and other shortages have significantly limited its ability to conduct large-scale combined-arms training exercises, such as those practised by US and South Korean forces. Fuel shortages have especially limited air force training: pilot training – according to anecdotal evidence – is limited to a handful of flying hours every year, because available aviation fuel needs to be conserved for actual military contingencies.

Nonetheless, despite shortages of spare parts, fuel and training time, North Korea's conventional capabilities pose a significant threat to allied forces and South Korea's population. For example, North Korea's

'On paper, North Korea's armed forces are formidable, but their actual capabilities are less than the raw data suggest, given the obsolescence of most North Korean equipment ... Nonetheless, North Korea's conventional capabilities pose a significant threat to allied forces and South Korea's population'

The Conventional Military Balance on the Korean Peninsula

Extracted from *The Military Balance 2003-2004*

Korea, Democratic People's Republic of (North) DPRK

Tinted text denotes information updated from previous year.

Total Armed Forces

ACTIVE ε1,082,000

Terms of service **Army** 5–12 years **Navy** 5–10 years **Air Force** 3–4 years, followed by compulsory part-time service to age 40. Thereafter service in the Worker/Peasant Red Guard to age 60

RESERVES 4,700,000 of which

Army 600,000 **Navy** 65,000 are assigned to units (see also *Paramilitary*)

Army ε950,000

20 Corps (1 armd, 4 mech, 12 inf, 2 arty, 1 capital defence)
• 27 inf div • 15 armd bde • 14 inf • 21 arty • 9 MRL bde
Special Purpose Forces Comd (88,000): 10 *Sniper* bde (incl 2 amph, 2 AB), 12 lt inf bde (incl 3 AB), 17 recce, 1 AB bn, 'Bureau of Reconnaissance SF' (8 bn)
Army tps: 6 hy arty bde (incl MRL), 1 *Scud* SSM bde, 1 FROG SSM regt
Corps tps: 14 arty bde incl 122mm, 152mm SP, MRL

RESERVES

40 inf div, 18 inf bde

EQUIPMENT

MBT some 3,500: T-34, T-54/-55, T-62, Type-59
LT TK 560 PT-76, M-1985
APC 2,500 BTR-40/-50/-60/-152, PRC Type-531, VTT-323 (M-1973), some BTR-80A
TOTAL ARTY (excl mor) 10,400
TOWED ARTY 3,500: **122mm:** M-1931/-37, D-74, D-30; **130mm:** M-46; **152mm:** M-1937, M-1938, M-1943
SP ARTY 4,400: **122mm:** M-1977, M-1981, M-1985, M-1991; **130mm:** M-1975, M-1981, M-1991; **152mm:** M-1974, M-1977; **170mm:** M-1978, M-1989
COMBINED GUN/MOR: 120mm (reported)
MRL 2,500: **107mm:** Type-63; **122mm:** BM-21, BM-11, M-1977/-1985/-1992/-1993; **240mm:** M-1985/-1989/-1991
MOR 7,500: **82mm:** M-37; **120mm:** M-43 (some SP); **160mm:** M-43
SSM 24 FROG-3/-5/-7; some 30 *Scud*-B/C (200+ msl), ε10 *No-dong* (ε90+ msl)
ATGW: AT-1 *Snapper*, AT-3 *Sagger* (some SP), AT-4 *Spigot*, AT-5 *Spandrel*
RCL 82mm: 1,700 B-10
AD GUNS 11,000: **14.5mm:** ZPU-1/-2/-4 SP, M-1984 SP; **23mm:** ZU-23, M-1992 SP; **37mm:** M-1939, M-1992; **57mm:** S-60, M-1985 SP; **85mm:** KS-12; **100mm:** KS-19

SAM ε10,000+ SA-7/-16

Navy ε46,000

BASES East Coast Toejo Dong (HQ), Changjon, Munchon, Songjon-pardo, Mugye-po, Mayang-do, Chaho Nodongjagu, Puam-Dong, Najin **West Coast** Nampo (HQ), Pipa Got, Sagon-ni, Chodo-ri, Koampo, Tasa-ri 2 Fleet HQ

SUBMARINES 26

SSK 26
22 PRC Type-031/FSU *Romeo* with 533mm TT, 4 FSU *Whiskey*† with 533mm and 406mm TT
(Plus some 45 SSI and 21 *Sang-O* SSC mainly used for SF ops, but some with 2 TT, all †)

PRINCIPAL SURFACE COMBATANTS 3

FRIGATES 3
FF 3
1 *Soho* with 4 SS-N-2 *Styx* SSM, 1 × 100mm gun and hel deck, 4 ASW RL
2 *Najin* with 2 SS-N-2 *Styx* SSM, 2 × 100mm guns, 2 × 5 ASW RL

PATROL AND COASTAL COMBATANTS some 310

CORVETTES 6
4 *Sariwon* FS with 1 × 85mm gun
2 *Tral* FS with 1 × 85mm gun
MISSILE CRAFT 43
15 *Soju*, 8 FSU *Osa*, 4 PRC *Huangfeng* PFM with 4 SS-N-2 *Styx* SSM, 6 *Sohung*, 10 FSU *Komar* PFM with 2 SS-N-2 *Styx* SSM
TORPEDO CRAFT some 103
3 FSU *Shershen* PFT with 4 × 533mm TT
60 *Ku Song* PHT
40 *Sin Hung* PHT
PATROL CRAFT 158
COASTAL 25
6 *Hainan* PFC with 4 ASW RL, 13 *Taechong* PFC with 2 ASW RL, 6 *Chong-Ju* with 1 × 85mm gun, (2 ASW mor)
INSHORE some 133
18 SO-1<, 12 *Shanghai* II<, 3 *Chodo*<, some 100<

MINE WARFARE 23

MINE COUNTERMEASURES about 23 MSI<

AMPHIBIOUS 10

10 *Hantae* LSM, capacity 350 tps, 3 tk
plus craft 15 LCM, 15 LCU, about 100 *Nampo* LCVP, plus about 130 hovercraft

SUPPORT AND MISCELLANEOUS 7

2 AT/F, 1 AS, 1 ocean and 3 inshore AGHS

COASTAL DEFENCE

2 SSM regt: *Silkworm* in 6 sites, and probably some mobile launchers
GUNS 122mm: M-1931/-37; **130mm:** SM-4-1, M-1992; **152mm:** M-1937

Air Force 86,000

4 air divs. 1st, 2nd and 3rd Air Divs (cbt) responsible for N, E and S air defence sectors respectively. 8th Air Div (trg) responsible for NE sector.
33 regts (11 ftr/fga, 2 bbr, 7 hel, 7 tpt, 6 trg) plus 3 indep air bns (recce/EW, test and evaluation, naval spt). The AF controls the national airline
Approx 70 full time/contingency air bases
605 cbt ac, ε24 armed hel
Flying hours 20 or less
BBR 3 lt regt with 80 H-5 (Il-28)
FGA/FTR 15 regt
    6 with 107 J-5 (MiG-17), 4 with 159 J-6 (MiG-19), 5 with 130 J-7 (MiG-21), 1 with 46 MiG-23, 1 with 30 MiG-29 (25 -A, 5 -U), 1 with 18 Su-7, 1 with 35 Su-25
TPT ac ε300 An-2/Y-5 (to infiltrate 2 air force sniper brigades deep into ROK rear areas), 6 An-24, 2 Il-18, 4 Il-62M, 2 Tu-134, 4 Tu-154
HEL 306. Large hel aslt force spearheaded by 24 Mi-24*. Tpt/utility: 80 Hughes 500D, 139 Mi-2, 15 Mi-8/-17, 48 Z-5
TRG incl 10 CJ-5, 7 CJ-6, 6 MiG-21, 170 Yak-18, 35 FT-2 (MiG-15UTI)
UAV Shmel
MISSILES
    AAM AA-2 *Atoll*, PL-5, PL-7, AA-7 *Apex*, AA-8 *Aphid*, AA-10 *Alamo*, AA-11 *Archer*
    SAM 19 SAM bde (40+ SA-2, 7 SA-3, 2 SA-5) with some 340 launchers/3,400 missiles, many thousands of SA-7/14/16. Possible W systems, reverse-engineered.

Forces Abroad

advisers in some 12 African countries

Paramilitary 189,000 active

SECURITY TROOPS (Ministry of Public Security) 189,000 incl border guards, public safety personnel

WORKER/PEASANT RED GUARD some 3,500,000 (R)
Org on a provincial/town/village basis; comd structure is bde – bn – coy – pl; small arms with some mor and AD guns (but many units unarmed)

Defence spending

won		2001	2002	2003
GNP	US$	ε18bn	ε20bn	
per capita	US$	804	903	
Growth	%	3.7		
Def exp	US$	ε4.5bn	ε5bn	
Def bdgt	won	2.9bn	3.2bn	3.6bn
	US$	1.3bn	1.4bn	1.6bn
US$1=won		2.2	2.2	2.2

Population			ε22,145,000	
Age	13–17	18–22	23–32
Men	1,074,000	908,000	2,504,000
Women	1,117,000	1,004,000	2,048,000

artillery capability does not require sophisticated tactics nor modes of operation to pose a threat to Seoul. In any conflict, North Korean artillery, firing from fortified positions near the DMZ, could initially deliver a heavy bombardment on the South Korean capital. Allied counter-battery fire and air strikes would eventually reduce North Korea's artillery capability, but not before significant damage and high casualties had been inflicted on Seoul. Similarly, the North Korean air force could launch surprise attacks against military and civilian targets throughout South Korea before allied air superiority was established. The potential delivery of chemical or biological weapons by artillery, short-range missiles and aerial bombs is an additional threat – especially to unprotected civilians.

North Korean naval forces could complicate US efforts to reinforce its forces in South Korea during a conflict. Although the US Office of Naval Intelligence states that North Korea's submarine force is obsolete and is 'only modestly proficient in basic operations in its own coastal waters', it points out that North Korean submarines could be effective in operations such as mining and insertion of Special Forces.[4] Similarly, although North Korea's missile and torpedo craft are old and their weapons systems obsolete, these vessels would need to be neutralised before US vessels could anchor and use ports to deliver reinforcements. Mines in the difficult waters around the Korean Peninsula would pose a threat to allied forces, as they did to the US 6th Fleet during the Korean War. However, the mine countermeasures capabilities of the US and South Korean Navies can now detect and clear mines at a much faster rate than previously.

United States Navy (USN)/United States Marine Corps (USMC) amphibious doctrine dictates that landings will be launched from relatively far out to sea, thereby keeping the large amphibious-capability vessels clear of shallow-water mines and coastal defence batteries. Moreover, although heavy equipment still relies on displacement landing vessels, a far greater proportion of amphibious troops and equipment are now transported by hovercraft and helicopter, which are less vulnerable to mines.

North Korea's military capability is also affected by issues of manpower effectiveness, including morale and loyalty. The political loyalty of officers and troops to Pyongyang is extremely difficult to determine, but party control and indoctrination remains strong in the armed forces, and there is no basis to assume that the military would collapse or revolt in wartime. As they demonstrated during the Korean War, North Korean forces are likely to be physically tough and resilient. However, years of maltreatment of soldiers by officers, and malnutrition – North Korean units rear livestock and grow vegetables to boost food stocks – may have affected morale to an unknown extent.

The Conventional Military Balance on the Korean Peninsula

South Korean military capabilities

South Korea's armed forces comprise approximately 686,000 active-duty troops and 4.5m reservists.[5] Its active ground forces are about half the size of North Korea's in terms of personnel, major equipment holdings and force structure, but its equipment is superior. South Korea's air and naval forces are comparable in size to North Korea's, and they possess much more modern and sophisticated equipment. Overall, South Korea's armed forces have become one of the world's more capable militaries and present a formidable forward defence against any possible attack by North Korea.

South Korea's army consists of 11 corps, with 52 divisions and 20 brigades. They can deploy some 2,300 main battle tanks, 2,500 armoured personnel carriers and light tanks, 4,500 heavy-calibre artillery pieces, 6,000 mortars, an estimated 600 air defence guns, over 1,000 surface-to-air missiles, and about a dozen short-range surface-to-surface missiles. Usually, 12 army divisions are deployed along the DMZ in heavily fortified positions. The South Korean air force has 538 combat aircraft and 117 attack helicopters. Meanwhile, the South Korean navy includes 39 principal surface combatants, 20 submarines, 84 patrol and coastal combatants, 15 mine warfare ships, 12 amphibious vessels, and 60 naval combat aircraft. South Korea's defence expenditure is several times more than that of North Korea. In 2002, as at average annual exchange rates, South Korea's defence budget amounted to $13.2bn. However, this figure needs to be balanced as manpower costs in the South are greater.

According to the TASCFORM scoring system, South Korea's ground combat weapon capabilities are rated higher than those of North Korea because of South Korea's qualitative edge. By the same measure, its air capabilities, when factoring in attack helicopters, are also superior – totalling about 2.5 F-16 wing equivalents. With the acquisition of the US Army Tactical Missile System (ATACMS) Block 1-A, due in service this year, South Korea's armed forces will increase their capabilities significantly. The missile system has a range of 300km and can target command and communications facilities, intelligence assets, and missile launching sites.

As measured by static equipment indices, South Korea's conventional forces would appear superior to North Korea's. When morale, training, equipment maintenance, logistics, and reconnaissance and communications capabilities are factored in, this qualitative advantage increases. In addition, if North Korea invaded the country, South Korean forces would have the advantage of fighting from prepared defensive positions. Therefore, the Pentagon's official current assessment of the Korean military balance suggests that, due to qualitative advantages, the South Korean–US combined force capabilities are superior to those of North Korea.[6]

Still, there are spheres in which South Korea could improve. A number of modernisation programmes have been delayed or cancelled due to financial considerations. According to the US, South Korea's principal shortcomings are in the areas of command, control and communications, chemical and biological defences, and precision munitions.[7] In addition, replacement of South Korea's ageing *Nike-Hercules* air defence system with *Patriot* missiles would significantly strengthen the country's theatre missile defence capability.

US military capabilities in Korea

A war in Korea would be extremely demanding on US forces. As envisaged in all post-Cold War defence plans (the 'base force' of President George H.W. Bush's administration, the Clinton administration's 1993 Bottom-Up Review and 1997 Quadrennial Defense Review, and the 2001 Quadrennial Defense Review of President George W. Bush's administration) the US would have to deploy almost half of its combat forces in the event of a fullscale conflict on the Korean Peninsula. The required number of troops would be in the region of 500,000–600,000, comparable to the 550,000 who fought in *Operation Desert Storm*.

The US normally has about 300 fixed-wing combat aircraft based in the immediate vicinity of the Korean Peninsula. This figure includes US air forces in Japan, and some 75 fixed-wing aircraft on board a carrier, which is normally deployed nearby. In March 2003, a number of F-117 stealth fighters and F-15E *Strike Eagle* fighters boosted US air capabilities when they were dispatched to the theatre for the annual *Foal Eagle* military exercise, but remained in the region because of tensions concerning North Korea's nuclear programme. A modest additional ground force element of perhaps 1,000 troops was also added at the same time, and 24 B-52 and B-1 bombers were deployed to Guam. These deployments demonstrate Washington's ability to quickly reinforce its military: the US could double its available combat aircraft in the region within a week, and double them again in another couple of weeks. Moreover, US Army and Marine Corps forces in northeast Asia have 100 attack helicopters deployed with them. Airfields available for US combat aircraft

'The Pentagon's official current assessment of the Korean military balance suggests that, due to qualitative advantages, South Korean–US combined force capabilities are superior to those of North Korea'

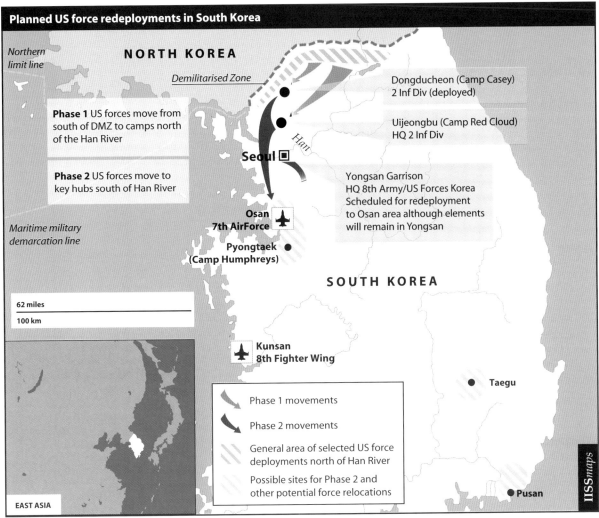

Planned US force redeployments in South Korea

Northern limit line

NORTH KOREA

Demilitarised Zone

Phase 1 US forces move from south of DMZ to camps north of the Han River

Phase 2 US forces move to key hubs south of Han River

Maritime military demarcation line

Han

Seoul ◼

Dongducheon (Camp Casey)
2 Inf Div (deployed)

Uijeongbu (Camp Red Cloud)
HQ 2 Inf Div

Yongsan Garrison
HQ 8th Army/US Forces Korea
Scheduled for redeployment
to Osan area although elements
will remain in Yongsan

Osan
7th AirForce ✈

Pyongtaek ●
(Camp Humphreys)

SOUTH KOREA

62 miles
100 km

Kunsan
8th Fighter Wing ✈

● **Taegu**

Phase 1 movements

Phase 2 movements

General area of selected US force
deployments north of Han River

Possible sites for Phase 2 and
other potential force relocations

● **Pusan**

EAST ASIA

IISS*maps*

Source: **American forces in South Korea: the end of an era? IISS** *Strategic Comments* **Volume 9. no 5. July 2003**

would number at least half a dozen at the start of any hostilities, and could quickly be expanded to a dozen or more locations across Japan and South Korea.

The US usually stations two brigades of the Army's Second Infantry Division in South Korea. These constitute 18,000 US troops, based in 17 camps approximately halfway between Seoul and the DMZ and astride the two main potential avenues of approach in the western half of the country. US forces conduct extensive training, both on their own and in conjunction with the South Korean military, to maintain capability and interoperability. By 2006, US forces will have redeployed 75km south of the Han River (which runs through Seoul) with the headquarters of US Forces Korea moving from Seoul to the Osan-Pyongtaek area. The redeployment will not result in any significant reduction in the numbers or readiness of US forces on the Peninsula; nor will it remove completely the 'tripwire' function of US forces

against any North Korean attack. The redeployment, will, however, shorten US lines of communication in the initial phase of any confrontation and increase the survivability of US units by placing them out of range of initial North Korean artillery strikes. In this respect, the redeployment of US forces in South Korea will strengthen US military capabilities to defend South Korea and respond to any North Korean attack.

Politically, however, the redeployment plan has created nervousness in both Seoul and Pyongyang because of concerns that the US will be more inclined to launch a pre-emptive strike on North Korean nuclear and missile facilities if US troops are less vulnerable to North Korean retaliation. However, the risk that a pre-emptive strike would lead to a general conflict is very high, regardless of whether or not US forces are immediately exposed to North Korean retaliation. Nonetheless, Pyongyang may respond to the US redeployment by seeking to enhance its existing

The Conventional Military Balance on the Korean Peninsula

Extracted from *The Military Balance 2003-2004*

Korea, Republic of (South) ROK

Tinted text denotes information updated from previous year.

Total Armed Forces

ACTIVE 686,000

(incl ε159,000 conscripts)
Terms of service conscription **Army** 26 months **Navy** and **Air Force** 30 months; First Combat Forces (Mobilisation Reserve Forces) or Regional Combat Forces (Homeland Defence Forces) to age 33

RESERVES 4,500,000

being re-org

Army 560,000

(incl 140,000 conscripts)
HQ: 3 Army, 11 Corps (two to be disbanded)
3 mech inf div (each 3 bde: 3 mech inf, 3 tk, 1 recce, 1 engr bn; 1 fd arty bde) • 19 inf div (each 3 inf regt, 1 recce, 1 tk, 1 engr bn; 1 arty regt (4 bn)) • 2 indep inf bde • 7 SF bde • 3 counter-infiltration bde • 3 SSM bn with NHK-I/-II (*Honest John*) • 3 AD arty bde • 3 I HAWK bn (24 sites), 2 *Nike Hercules* bn (10 sites) • 1 avn comd with 1 air aslt bde

RESERVES

1 Army HQ, 23 inf div

EQUIPMENT

MBT 1,000 Type 88, 80 T-80U, 400 M-47, 850 M-48
AIFV 40 BMP-3
APC incl 1,700 KIFV, 420 M-113, 140 M-577, 200 Fiat 6614/KM-900/-901, 20 BTR-80
TOWED ARTY some 3,500: **105mm:** 1,700 M-101, KH-178; **155mm:** M-53, M-114, KH-179; **203mm:** M-115
SP ARTY 155mm: 1,040 M-109A2, ε36 K-9; **175mm:** M-107; **203mm:** 13 M-110
MRL 130mm: 156 *Kooryong* (36-tube); **227mm:** 29 MLRS (all ATACMS capable)
MOR 6,000: **81mm:** KM-29; **107mm:** M-30
SSM 12 NHK-I/-II
ATGW TOW-2A, *Panzerfaust*, AT-7 *Saxhorn*
RCL 57mm, 75mm, 90mm: M67; **106mm:** M40A2
ATK GUNS 58: **76mm:** 8 M-18; **90mm:** 50 M-36 SP
AD GUNS 600: **20mm:** incl KIFV (AD variant), 60 M-167 *Vulcan*; **30mm:** 20 B1 HO SP; **35mm:** 20 GDF-003; **40mm:** 80 L60/70, M-1
SAM 350 *Javelin*, 60 *Redeye*, ε200 *Stinger*, 170 *Mistral*, SA-16, 110 I HAWK, 200 *Nike Hercules*, *Chun Ma* (reported)
SURV RASIT (veh, arty), AN/TPQ-36 (arty, mor), AN/TPQ-37 (arty)

HEL
ATTACK 60 AH-1F/-J, 45 Hughes 500 MD, 12 BO-105
TPT 18 CH-47D, 6 MH-47E
UTL 130 Hughes 500, 20 UH-1H, 130 UH-60P, 3 AS-332L

Navy 63,000

(incl 28,000 Marines; ε19,000 conscripts)
COMMANDS 1st Tonghae (Sea of Japan); **2nd** Pyongtaek (Yellow Sea); **3rd** Chinhae (Korean Strait)
BASES Chinhae (HQ), Cheju, Mokpo, Mukho, Pohang, Pusan, Pyongtaek, Tonghae

SUBMARINES 20

SSK 9 *Chang Bogo* (Ge T-209/1200) with 8 × 533 TT
SSI 11
3 KSS-1 *Dolgorae* (175t) with 2 × 406mm TT
8 *Dolphin* (175t) with 2 × 406mm TT

PRINCIPAL SURFACE COMBATANTS 39

DESTROYERS 6
DDG 6
3 *King Kwanggaeto* with 8 *Harpoon* SSM, 1 *Sea Sparrow* SAM, 1 × 127mm gun, 1 *Super Lynx* hel
3 *Kwang Ju* (US *Gearing*) with 2 × 4 *Harpoon* SSM, 2 × 2 × 127mm guns, 2 × 3 ASTT, 1 × 8 ASROC SUGW, 1 *Alouette* III hel
FRIGATES 9
FFG 9 *Ulsan* with 2 × 4 *Harpoon* SSM, 2 × 76mm gun, 2 × 3 ASTT (Mk 46 LWT)
CORVETTES 24
24 *Po Hang* FS with 2 × 3 ASTT; some with 2 × 1 MM-38 *Exocet* SSM

PATROL AND COASTAL COMBATANTS 84

CORVETTES 4 *Dong Hae* FS with 2 × 3 ASTT
MISSILE CRAFT 5
5 *Pae Ku*-52 (US *Asheville*) PFM, 2 × 2 *Harpoon* SSM, 1 × 76mm gun
PATROL, INSHORE 75
75 *Kilurki*-11 (*Sea Dolphin*) 37m PFI

MINE WARFARE 15

MINELAYERS 1
1 *Won San* ML
MINE COUNTERMEASURES 14
6 *Kan Keong* (mod It *Lerici*) MHC
8 *Kum San* (US MSC-268/289) MSC

AMPHIBIOUS 12

4 *Alligator* (RF) LST, capacity 700
6 *Un Bong* (US LST-511) LST, capacity 200 tps, 16 tk
2 *Ko Mun* (US LSM-1) LSM, capacity 50 tps, 4 tk
Plus about 36 craft; 6 LCT, 10 LCM, about 20 LCVP

SUPPORT AND MISCELLANEOUS 14

3 AOE, 2 spt AK, 2 AT/F, 2 salv/div spt, 1 ASR, about 4 AGHS (civil-manned, Ministry of Transport-funded)

NAVAL AVIATION

EQUIPMENT
16 cbt ac; 43 armed hel
AIRCRAFT
ASW 8 S-2E, 8 P-3C *Orion*
MR 5 *Cessna* F406
HELICOPTERS
ASW 22 MD 500MD, 10 SA 316 *Alouette* III, 11 *Lynx* Mk 99
UTL 2 206B *Jetranger*

MARINES (28,000)

2 div, 1 bde • spt units

EQUIPMENT
MBT 60 M-47
AAV 60 LVTP-7, 42 AAV-7A1
TOWED ARTY 105mm, 155mm
SSM *Harpoon* (truck-mounted)

Air Force 63,000

4 Cmds (Ops, Southern Combat, Logs, Trg), Tac Airlift Wg and Composite Wg are all responsible to ROK Air Force HQ. 538 cbt ac, no armed hel
FTR/FGA 7 tac ftr wgs
2 with 153 F-16C/D (104 -C, 49 -D)
3 with 185 F-5E/F (150 -E, 35 -F)
2 with 130 F-4D/E (60 -D, 70 -E)
CCT 1 wg with 22* A-37B
FAC 1 wg with 20 O-1A, 10 O-2A
RECCE 1 gp with 18* RF-4C, 5* RF-5A, 3 Hawker 800RA
ELINT/SIGINT 7 Hawker 800XP
SAR 1 hel sqn, 5 UH-1H, 4 Bell-212
TAC AIRLIFT WG ac 2 BAe 748 (VIP), 1 Boeing 737-300 (VIP), 1 C-118, 10 C-130H, 20 CN-235M/-220 **hel** 6 CH-47, 3 AS-332, 3 VH-60
TRG 25* F-5B, 50 T-37, 30 T-38, 25 T-41B, 18 *Hawk* Mk-67
UAV 3 *Searcher*, 100 *Harpy*
MISSILES
ASM AGM-65A *Maverick*, AGM-88 HARM, AGM-130, AGM-142
AAM AIM-7 *Sparrow*, AIM-9 *Sidewinder*, AIM-120B AMRAAM

Forces Abroad

KYRGYZSTAN (OP ENDURING FREEDOM): 90 (medical staff)

UN AND PEACEKEEPING

CYPRUS (UNFICYP): 1 **EAST TIMOR** (UNMISET): 225 **GEORGIA** (UNOMIG): 7 obs **INDIA/PAKISTAN** (UNMOGIP): 9 obs

Paramilitary ε4,500 active

CIVILIAN DEFENCE CORPS 3,500,000 (R) (to age 50)

MARITIME POLICE ε4,500
PATROL CRAFT 81
OFFSHORE 10
3 *Mazinger* (HDP-1000) (1 CG flagship), 1 *Han Kang* (HDC-1150), 6 *Sea Dragon/Whale* (HDP-600)
COASTAL 33
22 *Sea Wolf/Shark*, 2 *Bukhansan*, 7 *Hyundai*-type, 2 *Bukhansan*
INSHORE 38
18 *Seagull*, about 20<, plus numerous boats
SUPPORT AND MISCELLANEOUS 3 salvage
HEL 9 Hughes 500

Foreign Forces

US 38,500: **Army** 28,600; 1 Army HQ, 1 inf div **Navy** 300 **Air Force** 9,420: 1 HQ (7th Air Force); 84 cbt ac, 2 ftr wg; 3 sqn with 60 F-16, 1 sqn with 12 A-10, 12 OA-10, 1 special ops sqn **USMC** 180

Defence spending

won		2001	2002	2003
GDP	won	545tr	596tr	
	US$	422bn	476bn	
per capita	US$	8,970	10,035	
Growth	%	3.0	6.3	
Inflation	%	4.0	2.7	
Debt	%	17.2	16.2	
Def exp	won	14.7tr	16.7tr	
	US$	11.4bn	13.3bn	
Def bdgt	won	15.3tr	16.5tr	17.9tr
	US$	11.9bn	13.2bn	14.8bn
US$1=won		1,290	1,251	1,215
Population				**47,479,000**
Age	13–17	18–22	23–32
Men	1,780,000	1,916,000	4,359,000
Women	1,672,000	1,784,000	4,088,000

South Korean K-1 MBT

capabilities to attack targets beyond standard artillery range – by expanding its short-range ballistic missile force, for example.

In an emergency, US ground forces in Korea can be roughly tripled in size within ten days. Initial reinforcements would include the 25th Infantry Division from Hawaii. In addition, a brigade's worth of army equipment and a brigade's worth of Marine Corps equipment stored in pre-positioned ships in the Indian Ocean would arrive shortly thereafter, to be manned by troops airlifted from the US. After several weeks, a number of ships could also arrive from the US. Eight SL-7 fast sealift ships carrying a US-based heavy armoured army division could reach Korea after some 20–30 days. In the same timeframe, many large, medium-speed, roll-on/roll-off vessels, as well as more ground forces and marines could also reach the Peninsula. More aircraft carriers and other ships, possibly serving in the Mediterranean, Persian Gulf, or off the west coast of the US could also be redeployed. Within 75 days, according to official plans, the entire transport operation could be complete. In practice, the operation might take 100 days given the inevitable complications concerning actual deployments and the potential need to clear North Korean mines, submarines, and missile boats from South Korean waters before unloading supply ships.

In all, using the TASCFORM system, these US forces would correspond to at least five modern heavy ground-division equivalents and more than 15 modern fighter wings – a combined capability exceeding that of North Korea's forces, even without including South Korea in the balance. However, the complex logistics of moving large numbers of forces and equipment over great distances, potentially in the face of North Korean countermeasures, would create a difficult and demanding task. Moreover, in the time it takes for full reinforcements to arrive and be deployed in the region – at least 2–3 months – the military situation could have altered dramatically.

Finally, in extremis, US forces would also be capable of using nuclear weapons in any Korean conflict. Although President George H.W. Bush ordered the unilateral removal of all tactical nuclear weapons from Korea in 1991, the US retains the ability to deploy nuclear-armed submarine launched cruise missiles (SLCMs) in response to any use of unconventional weapons by North Korea against US forces or allies. The US official negative security assurance concerning the threat or use of nuclear weapons against non-nuclear weapon states has not applied to North Korea

since 1993, when it was found to be in violation of its nuclear Non-Proliferation Treaty (NPT) safeguards obligations. In any event, Pyongyang has subsequently withdrawn from the NPT and declared that it possesses a 'nuclear deterrent force'. Moreover, longstanding US doctrine does not preclude the possible use of nuclear weapons against an adversary that uses chemical or biological weapons against the US or its allies, such as South Korea and Japan. North Korea is fully aware of US nuclear capabilities and has often cited the US nuclear threat as a justification for its own nuclear programme.

Military analysis of a theoretical North Korean invasion

Throughout the 1990s, US military planners assumed that a North Korean surprise attack on South Korean and US forces might succeed, at least in terms of achieving its intermediate objective of seizing Seoul. In order to accomplish this goal, North Korea would employ its two main advantages: mass artillery – already deployed within range of Seoul – and heavy armoured forces, deployed immediately north of the DMZ. In theory, a massive artillery barrage would stun defences and open corridors for armoured forces to punch through and capture Seoul before allied forces could react. In the event that coalition forces lost Seoul, US military planners estimate that approximately six US ground combat divisions including marine and army units, ten air force wings, and four to five carrier battle groups, would be required to liberate the South Korean capital.

In recent years, with the steady degradation of North Korean forces and improvements in South Korean military capabilities, some experts believe that the prospect of a successful North Korean attack to capture Seoul and reunify the Peninsula through force has diminished.[8] In this analysis, South Korea and US forces deployed along the DMZ and around Seoul would stand a good chance of halting or at least severely delaying a North Korean offensive. As a war progressed, North Korean forces would become increasingly vulnerable to US reinforcements, and to attack by precision munitions. Their supply lines would be disrupted by allied artillery and air attacks – which would interdict reinforcement and re-supply along predictable axes of advance.

In addition, South Korean defences are well-prepared, and the country's armed forces are qualitatively superior (while having comparable firepower to those of North Korea). With most of the

'Longstanding US doctrine does not preclude the possible use of nuclear weapons against an adversary that uses chemical or biological weapons against the US or its allies, such as South Korea and Japan'

South Korean army deployed across a 250km front (in a dense force-to-space ratio of about one division per 10km), invading forces operating in a hostile air environment would find it hard to penetrate such defences, particularly across terrain that is largely unsuitable for the movement of armour. The natural obstacles of rivers and marshes, combined with man-made barriers, mines, and bridge demolitions, would impede further movement and channel North Korean forces into killing zones.

Even assuming an attack in winter, which would allow forces to traverse frozen rice fields, North Korean military bridging units would need to cross the Han or Imjin Rivers in any attack on Seoul from the western half of the country. The combined effects of terrain, demolitions, and allied air and artillery in slowing North Korean armoured vehicles would greatly increase their vulnerability to direct fire from anti-tank systems and armour, as well as indirect fire and air-delivered precision munitions. To decrease the impact of allied strikes, North Korean forces might attempt to obstruct the visibility of US and South Korean forces by the use of artillery smoke rounds. Even if they lost some of their sensor and optic capability, allied forces would still have the advantage of being able to fire from protected positions.

Although South Korea's K-1 tanks do not have the detection and targeting capability of the US M1 *Abrams*, the K-1 is superior to and better protected than North Korea's outdated Soviet-type tanks and should prevail in tank versus tank engagements. The all-weather, day-night capability of the allied forces is another 'force multiplier', allowing troops to detect any massing of armoured vehicles with various platforms. This equipment includes reconnaissance satellites, RC-7B planes, and joint surveillance target attack radar system (JSTARS) aircraft, as well as ground radars and infrared intruder detection systems.

In the air, US and South Korean aircraft would quickly establish air superiority, with at least 500 planes and helicopters deployed to thwart a ground attack. Using 1991 Gulf War data as a guide, allied forces could expect to destroy roughly one armoured vehicle for every four shots fired, deploying *Maverick, Hellfire*, and tube-launched optically-tracked wire-guided (TOW) missiles as well as laser-guided bombs. Therefore, in theory, allied forces could destroy several hundred North Korean armoured vehicles per day, although this estimate could be affected by weather conditions and by the relative ease of concealment afforded by the Korean terrain, in comparison to the open desert of Kuwait and Iraq. Over time, North Korean targets would become fewer in number and more dispersed, but the rapid arrival of US air reinforcements would increase the density of airpower in the area of operations.

North Korean commando forces would only have a limited capability to disrupt South Korean defences. Firstly, an airborne assault generally requires air superiority and the suppression of the enemy's artillery and air defences – North Korea would not be able to gain either of those advantages. Secondly, a tunnel assault could be more effective, although troops arriving via underground passageways would be unable to penetrate deep into South Korean defences, given the short length of the tunnels. Also, they would be highly vulnerable to counter-attacks on the tunnel entrances by artillery or munitions delivered by air, which could probably be initiated within a short time of the assault starting. Thirdly, it is likely that North Korean Special Forces would use submarines to infiltrate troops, although their numbers would be limited. Nevertheless, Special Forces might be able to cause serious disruption – including through the possible delivery of chemical or biological agents – in cities and rearguard military areas.[9]

Finally, North Korea's economic decline continues to erode the relative effectiveness and readiness of its forces. Despite the regime's 'military first' policy, North Korea cannot afford to significantly modernise its ageing conventional forces – or even afford the levels of maintenance, refurbishment, and training necessary to maintain high readiness. In conclusion, North Korea is capable of inflicting widespread damage on South Korea in the early days of any conflict, but there is considerable doubt about the ability of North Korea's conventional forces to sustain offensive military operations and resist the counter-attacks of a technologically superior adversary with better trained forces. Therefore, according to the Commander of US Forces in Korea, General Schwartz, 'an attack scenario appears unlikely at this time because North Korea clearly knows that its regime would ultimately be destroyed as a result of any attack.'[10]

A US Pre-emption Option?

If North Korea cannot be confident of achieving a successful invasion of South Korea, do Washington and Seoul have a plausible offensive option to invade North Korea and capture Pyongyang? Despite the relative shift in the military balance of power in favour of the allies, such a pre-emptive use of force would appear very risky. Even though US and South Korean forces

'Even with the US military's prowess and ability to execute 'effects-based operations' (which aim to disrupt the decision-making ability of an adversary) an invasion of North Korea would likely prove much more costly than the 2003 Iraq War'

enjoy qualitative superiority, and, via a US military build-up, could increase this superiority rapidly, they could not be confident of winning an offensive war against North Korea without sustaining heavy military and collateral casualties. An all-out invasion, along the lines of the March–April 2003 campaign in Iraq, is not an appealing option. Even with the US military's prowess and ability to execute 'effects-based operations' (which aim to disrupt the decision-making ability of an adversary) an invasion of North Korea would likely prove much more costly than the 2003 Iraq War.

Firstly, with so many North Korean weapons deployed near Seoul, and many in protected locations, even a well-timed surprise attack could not prevent a heavy artillery bombardment of the South Korean capital. In their current positions, a large percentage of North Korea's estimated 10,000 or so artillery pieces are deployed within range of Seoul, with the capability to fire several rounds a minute. The initial speed of a fired shell is generally around half a kilometre per second. Therefore, even if an allied counter-battery radar, some 10km away picked up a North Korean missile or artillery shell and established a track on it within seconds, a counter-strike would not be able to silence the North Korean gun or launcher for at least a minute. As a result, one artillery piece could probably fire 2–5 rounds before being neutralised or forced to retreat into its shelter. Theoretically, several thousand artillery rounds could land in Seoul no matter how hard the allies tried to prevent or stop the attack.

Secondly, many North Korean military and political leadership facilities are located deep underground, making them hard to identify and attack, and thus limiting the effectiveness of effects-based operations. North Korea has studied US campaigns in the Balkans and Middle East and has taken measures – such as hardening facilities, dispersing forces, and improving camouflage, concealment and deception – to counter US technological advantages. Compared with Iraq, North Korea has more numerous hardened underground facilities, and the US has even worse intelligence on the functions or even the locations of such facilities. It would be extremely difficult for US Special Forces to infiltrate North Korea, to locate command locations and to direct aerial attacks against such facilities.

Thirdly, there is no straightforward line of approach to Pyongyang, like the open desert traversed by coalition forces in their advance on Baghdad in March–April 2003. Just as the difficult terrain would complicate a North Korean attack on the South, it would also present difficulties for any allied attack against the North. As a result, the harassment of supply lines during an invasion of North Korea could be a much more pervasive problem than it was in Iraq. An amphibious landing to bypass the DMZ and attack Pyongyang directly would be a substantial and risky operation, especially since the North has made efforts to strengthen its coastal defences in recent years. Airborne troops would presumably play an important role in seizing key assets behind North Korean lines, but it could be difficult to reinforce these troops if advances over land were stalled.

Fourthly, North Korea's armed forces, with a total active-duty strength of more than a million, are much larger than Iraq's and more likely to resist. Iraq's forces numbered around 400,000 (against coalition forces of more than 250,000) and most of Iraq's forces melted away under the coalition's overwhelming firepower. In contrast, the US military respects the toughness and determination of the individual Korean soldier. While difficult to measure, the North Korean regime probably commands more political legitimacy and loyalty than the Ba'athist regime in Iraq, and a collapse of North Korea's armed forces cannot be assumed. Consequently, an offensive operation against North Korea would require a build-up of hundreds of thousands of US troops, in addition to a large number of South Korean soldiers. It would be impossible to conceal this build up from Pyongyang, which would have time to prepare its defences or even launch its own pre-emptive attack before allied forces could be fully deployed.

Fifthly, North Korea's arsenal of unconventional weapons and ballistic missiles is estimated to be much more potent than Iraq's. In the case of Iraq, US military planners assumed that Baghdad had a relatively small arsenal of chemical and biological weapons, which would not be effective on the battlefield against protected coalition forces and could not be delivered in large amounts against cities beyond Iraq's borders. In contrast, North Korea is assumed to have thousands of tonnes of chemical and possibly biological weapons, which can be delivered against Seoul via numerous short-range weapons or against cities throughout South Korea and Japan via longer-range missiles. Iraq was thought to have a handful of ageing ballistic missiles, while North Korea is believed to have hundreds of ballistic missiles. Finally, Washington was confident that Iraq did not have any nuclear weapons, while North Korea is thought to possess a small nuclear capability. Though the use of its presumed nuclear arsenal would invite nuclear retaliation, Pyongyang has effectively tried to foster the impression that it would

'A full-scale pre-emptive attack to remove the North Korean regime is considered by Washington, Seoul and Tokyo to be an impractical option'

take suicidal actions as a last resort if faced with a military threat to extinguish its regime.

Due to these considerations, a fullscale pre-emptive attack to remove the North Korean regime is considered by Washington, Seoul and Tokyo to be an impractical option. More limited pre-emptive options include air strikes against known or suspected North Korean nuclear facilities, chemical weapons storage sites, missile launchers and firing bunkers, or North Korean artillery locations near the DMZ. But, these limited options suffer two basic disadvantages. Firstly, from a practical standpoint, it would be difficult to conduct a fully effective first strike, given the uncertainty and multiplicity of targets. For example, US forces could probably destroy the key nuclear facilities at the Yongbyon Nuclear Research Centre – thus destroying North Korea's known plutonium production capacity – but such attacks cannot account for additional facilities and materials presumed to be hidden at other locations. Similarly, pre-emptive attacks could only account for a portion of mobile missile launchers and artillery rockets.

Secondly, a limited pre-emptive attack runs the risk of provoking North Korean retaliation. Pyongyang would view any attack against its strategic assets as the opening move in a broader campaign to destroy the regime. On the one hand, Pyongyang will presumably recognise that a massive response against South Korea and Japan to a limited US strike would guarantee the end of the regime. On the other hand, it would be extremely difficult for Pyongyang to suffer a loss of key assets without responding in some way to defend itself and deter additional allied attacks. Even if Pyongyang tried to limit its response to selected targets, such as targeting US military forces deployed in South Korea, the danger of escalation to a state of general conflict would be high. As a consequence, there is little enthusiasm in Washington, and much less in Seoul and Tokyo, for a surprise 'surgical strike' to knock out North Korea's key military assets. However, if the allies believed that war was inevitable and that North Korea

was preparing to attack, a pre-emptive strike would hold great advantages. Likewise, if Pyongyang feared an attack on its critical military assets, it would be under pressure to use its weaponry before these assets could be destroyed on the ground.

Conclusion

The combination of North Korea's long economic decline and enhanced US and South Korean military capabilities has diminshed the threat of a North Korean invasion of South Korea. Nonetheless, North Korea retains the ability to inflict heavy casualties and collateral damage, largely through the use of massed artillery. In effect, Pyongyang has more of a threat to devastate Seoul than to seize and hold it. North Korea's conventional threat is also sufficient to make an allied pre-emptive invasion to overthrow the North Korean regime a highly unattractive option. In theory, US forces could carry out pre-emptive attacks to destroy known North Korean nuclear facilities and missile emplacements, but such attacks could provoke North Korean retaliation and trigger a general conflict.

North Korea cannot invade the South without inviting a fatal counter-attack from the US and South Korea, while Washington and Seoul cannot overthrow the North Korean regime by force or destroy its strategic military assets without risking devastating losses in the process. In this respect, the balance of forces that emerged from the Korean War, and which helped in maintaining the armistice for 50 years, remains in place. None of the principal parties want to fight a war although they are prepared to fight if necessary. In this respect, the balance of forces creates certain vulnerabilities since it places a high premium on carrying out a pre-emptive strike if one side or the other believes that an attack is imminent. The danger is that war will begin out of miscalculation, misperception and escalation, rather than design. As a consequence, reduction of political tensions and conventional confidence-building measures can help to reduce the risk of war.

'North Korea cannot invade the South without inviting a fatal counter-attack from the US and South Korea, while Washington and Seoul cannot overthrow the North Korean regime by force or destroy its strategic military assets without risking devastating losses in the process.'

The Conventional Military Balance on the Korean Peninsula

CONCLUSION

Conclusion

These are the primary conclusions of our assessment of North Korea's nuclear, chemical and biological weapons programmes, its ballistic missile programme and the conventional balance of forces on the Korean Peninsula.

Nuclear

Compared to other weapons programmes, North Korea's nuclear weapons efforts have received the lion's share of intelligence scrutiny and attention, especially by the US, which views nuclear proliferation as its highest priority issue with North Korea. Because of this attention, and IAEA inspections of several key facilities, more is known about North Korea's nuclear capabilities than its other weapons programmes, but many uncertainties remain. The two key questions affecting an evaluation of North Korea's nuclear weapons capabilities are: how much weapons-usable nuclear material (separated plutonium or highly enriched uranium) does North Korea have on hand and how much can it produce in the future; and is North Korea able to design and fabricate deliverable nuclear weapons with this fissile material?

- It is plausible that North Korea was able to produce enough plutonium before 1992 for one or possibly two nuclear weapons.

This judgement is based on a mixture of technical and motivational factors. Firstly, there is strong technical evidence of discrepancies between IAEA samples taken at the Yongbyon reprocessing facility and samples of the small amount of plutonium declared by Pyongyang in 1992, suggesting that North Korea actually separated more plutonium before 1992 than it declared to the IAEA. However, the amount of additional plutonium cannot be determined by these sampling discrepancies. Secondly, in theory, North Korea could have operated its Soviet-supplied IRT reactor and the 5MW(e) graphite-moderated reactor before 1992 to produce additional undeclared spent fuel containing a maximum of 8–12 kilograms of plutonium. In theory, this is enough plutonium for one or possibly two early generation implosion-type nuclear weapons, taking into consideration expected reprocessing losses (10–30%) and the amount of plutonium typically required for a simple fission weapon (5–8kg). Thirdly, the fact that North Korea sought to conceal suspect nuclear waste sites before and during the IAEA inspections in 1992 and then ran the risk of provoking an international crisis by refusing to allow access to these sites – which could have resolved questions about its pre-1992 plutonium production – suggests that Pyongyang was hiding something of value. Finally, it is plausible that having invested considerable time and resources to build indigenous plutonium production

facilities, Pyongyang would try to get some strategic benefit from its investment before turning the facilities over to IAEA inspection. In this scenario, Pyongyang's mistake was getting caught.

Despite the plausibility of this judgement, it may be wrong. There is no conclusive proof of how much plutonium North Korea actually produced and separated prior to 1992. The amount could be considerably less than the theoretical maximum of 8–12kg. Perhaps Pyongyang salted away a much smaller amount of plutonium – maybe a few kilograms – and planned to add to this cache by secretly diverting small amounts of fresh plutonium produced under IAEA safeguards, until it built up enough for a nuclear arsenal. In this scenario, North Korea's objective in the 1993–94 nuclear crisis was to preserve strategic ambiguity. As long as Washington believed that North Korea could have enough plutonium for one or two nuclear weapons – an assessment that was made public at the time – Pyongyang would achieve some degree of nuclear deterrence. Strategic ambiguity was maintained under the Agreed Framework, which froze additional plutonium production, but allowed North Korea to retain a presumed nuclear capacity for a period of years before it was required to account for this material.

- Aside from plutonium produced prior to 1992, North Korea clearly has enough additional plutonium on hand for a few nuclear weapons. Its ability to produce fresh plutonium over the next several years is limited, but could substantially expand by the end of the decade.

North Korea's most important strategic asset is the plutonium contained in nearly 8,000 spent fuel rods discharged from the 5MW(e) reactor in June 1994. The exact amount of plutonium contained in these fuel rods is unknown, but the IAEA estimates they contain about 25–30kg of plutonium – which is a reasonable estimate. Assuming that 10–30% of this plutonium is lost in reprocessing and assuming that 5–8kg of plutonium is required for a simple implosion weapon, this is enough for two to five nuclear weapons. Since North Korea broke the nuclear freeze in late 2002, the status of these rods – and the plutonium contained within – is unknown. North Korea claims that it began to reprocess the fuel in April 2003 and finished the job in July 2003 – a time frame consistent with the technical parameters of the reprocessing facility. During this period, satellite intelligence and environmental sampling detected indications that some reprocessing had begun, but the available information cannot verify North Korea's claim that it completed the work. Certainly, it is plausible that Pyongyang would take advantage of the 2003 Iraq War to extract the

Conclusion

plutonium, while US military resources were occupied elsewhere, but it is also possible that North Korea ran into technical difficulties or that private warnings from Washington and Beijing halted the operation in mid-stream. Assuming that some or all of the rods have been reprocessed, it is not possible to determine what has become of the extracted plutonium without on-site inspection. Extracted plutonium could be fabricated into nuclear weapon components, a process involving melting, casting and machining plutonium metal to form a spherical core. Or, North Korea could retain the plutonium in raw form (either as an oxide or metal), threatening to fabricate weapon components if concessions are not forthcoming.

Aside from the plutonium in the 8,000 spent fuel rods, North Korea's ability to produce fresh plutonium in the near term is limited. Operating at maximum power for 300 days a year, the 5MW(e) reactor is theoretically capable of producing 7.5kg of plutonium annually, enough for about one nuclear weapon, assuming a 10–30% reprocessing loss and 5–8kg of plutonium per weapon. Given its poor operating history, however, and the fact that it has been mothballed for nearly a decade, the reactor may have experienced start-up problems since North Korea declared that it had resumed operations in February 2003. Thus, assuming enough plutonium for one or two nuclear weapons separated before 1992 and enough plutonium for two to five nuclear weapons in existing spent fuel, as well as about one additional bomb's worth of plutonium produced by the 5MW(e) reactor annually, North Korea's maximum potential nuclear arsenal over the next several years is likely to be around half a dozen to a dozen nuclear weapons – if no new facilities to produce plutonium or highly enriched uranium come online. (This estimate assumes no access to foreign sources of weapons-usable nuclear materials.)

In the longer term, North Korea's ability to produce significantly larger quantities of plutonium depends on finishing the 50MW(e) reactor, which was thought to be one to two years from completion at the time of the 1994 freeze. However, the status of key pieces of equipment for the reactor has never been determined, and it is not known whether enough graphite or fuel was produced prior to the freeze to outfit the reactor. Fuel fabrication could be a bottleneck because parts of the existing fuel fabrication plant at Yongbyon are heavily corroded and would need to be rebuilt before fuel fabrication could resume. Assuming no major technical problems, the 50MW(e) reactor could probably be completed, tested and brought up to high power operations in a few years at the earliest, although the end of the decade is a more cautious prediction. If Pyongyang makes a political decision to complete the reactor, large-scale activity at the facility would be liable to detection by satellite intelligence

resources. There have been no public reports of such activity since North Korea announced, in December 2002, its intention to resume construction of the reactor. In theory, operating at full power for 300 days per year, the 50MW(e) reactor could produce about 55kg of plutonium each year in its spent fuel, enough for about 5–10 nuclear weapons – depending on reprocessing losses and the amount of plutonium required for each weapon of North Korean design. The much larger 200MW(e) reactor was not close to completion in 1994 and has suffered from poor maintenance during the freeze. It is many years away from completion, at best.

- There is convincing evidence that North Korea has embarked on a clandestine enrichment programme, but not enough information to determine with confidence the programme's status and when it might be completed.

The June 2002 US assessment that North Korea is seeking to develop a gas centrifuge plant capable of enriching enough weapons-grade uranium for 'two or more nuclear weapons a year' is based on several pieces of evidence. Firstly, there were indications that Pakistan provided centrifuge technology to North Korea in exchange for *No-dong* missiles in the late 1990s. While not conclusive, the evidence was plausible. Pakistan was presumably interested in obtaining 'off-the-shelf' intermediate-range missiles capable of delivering nuclear payloads, a task for which the *No-dong* is well-suited. Meanwhile, North Korea was presumably interested in obtaining an alternative to plutonium production to maintain its nuclear hedge, since the Agreed Framework would have required North Korea eventually to declare its plutonium stocks and dismantle its plutonium production facilities. Disclosures that Pakistani scientists may have sold centrifuge technology to Iran and Libya increases the plausibility of a bargain with North Korea over missiles for nuclear technology. Secondly, according to press reports, South Korea and the US obtained information in 2001 about the centrifuge programme from a North Korean source, said to be a defector with some knowledge of the programme. From public information, it is impossible to make an independent judgement of the reliability of this source, but the information was apparently considered credible enough by US and South Korean intelligence agencies to be taken seriously.

Thirdly – and perhaps most convincing – North Korean procurement attempts strongly point to an effort to acquire materials and equipment for a production-scale centrifuge plant. Some of this information is public, most notably the April 2003 interdiction of a shipment of 22 tonnes of high-strength aluminium tubes – the first installment of a 200 tonne order. The particular type of aluminium and the dimensions of the tubes (unlike the tubes that Iraq was trying to acquire before the 2003 war) closely match the requirements for rotor casings for known centrifuge designs – in this case a type of centrifuge known to be in Pakistan's possession. Allowing for processing losses, the 200 tonnes of tubes could theoretically be used in the manufacture of about 3,500 of these centrifuge machines, enough to produce about 75kg of weapons-grade uranium a year or roughly three nuclear weapons of a first generation uranium-based implosion design (assuming 20–25kg per weapon). There is, apparently, additional information on procurement activities that has not been made public. For American officials, Vice Minister Kang's 'admission' to Secretary Kelly in October 2003 served to reinforce their conclusion that North Korea had an enrichment programme. Since then, North Korea has said that Kang was 'misunderstood' and has denied that it has an enrichment programme. However, the US conclusion is not based on the statements of North Korean officials one way or the other.

Publicly, the US estimates that a production-scale centrifuge facility 'could be operational as soon as mid-decade.' However, it is difficult to reach a confident assessment because many key factors are unknown, such as: the extent of the assistance provided by Pakistan; whether North Korea has been able to obtain all of the equipment and materials necessary to complete such a facility; the extent to which it can produce such items indigenously; and, most importantly, the location and status of the centrifuge plant and ancillary facilities. Of course, if all goes well, it is possible that the plant 'could' be finished by mid-decade, but plant completion and operation could also be delayed by interdiction efforts, such as the April 2003 tube seizure, and technical difficulties typically experienced by centrifuge programmes. From publicly available information, it is not possible to make a firm judgement.

- Assuming that it has sufficient fissile materials, it is plausible that North Korea could design and fabricate a simple implosion device, based on either plutonium or highly enriched uranium. But there is not enough information to reach firm conclusions about the details of such a weapon and how the North Koreans could deliver such a payload.

Given North Korea's long history of nuclear-related high-explosive testing, which began in the mid-1980s and continues through to the present, it is reasonable to assume that North Korea has been able to master the technology for a simple implosion device by now – the basics of the technology are widely known. North Korea is thought to possess the necessary skills in fields such as physics, electronics, munitions production and

metallurgy, to design and fabricate such a device. Moreover, high-explosive testing – with surrogate materials in place of a fissile core – can be used to develop a reliable design without the need for a full nuclear test. Based on these considerations, Russian and US intelligence agencies have, since the early to mid-1990s, assessed that North Korea is capable of building 'simple fission-type' nuclear weapons without the need to conduct nuclear tests. This judgement has become more confident over time. The political assumption is that North Korea would build a bomb if it could. Certainly, Pyongyang wants the outside world to believe that it has such weapons.

The technical details of a presumed North Korean nuclear weapon are unknown. Physics dictates that all simple implosion weapons share certain characteristics, but there are wide variations in overall size and weight. A key question is whether North Korea can build a nuclear weapon small and light enough to be delivered by the *No-dong* missile. Clearly, from Pyongyang's standpoint, it would be desirable to develop a nuclear warhead for the *No-dong* missile – which would be a more reliable and effective delivery system than aircraft and therefore enhance deterrence. In this regard, a critical factor yet to be determined is whether, or to what extent, Pakistan may have provided nuclear weapons design information or even weapons-grade uranium to North Korea as part of the missile-for-nuclear deal. Of course, it is not known whether North Korea actually has deliverable nuclear weapons, but it would be imprudent to assume that it does not.

Chemical and biological weapons
Compared to its nuclear programme, far less is known about North Korea's chemical and biological weapons (CBW) capabilities. In general, CBW programmes are more difficult for intelligence agencies to detect and monitor than nuclear programmes because of the nature of the technology and facilities involved. In contrast to North Korea's nuclear facilities at Yongbyon, North Korea's dual-use civilian chemical and biological infrastructure has never been subject to international inspection, much less the suspect facilities reportedly associated with chemical or biological weapons research, production and storage. The US has generally placed a higher priority on gathering and analysing information on North Korea's nuclear programme, with the result that fewer US intelligence resources have been devoted to CBW activities. South Korea, however, has seen the potential CBW threat as a more important intelligence priority, and its public government reports have included more detail than comparable reports from the US and Russia.

- North Korea has probably produced and stockpiled chemical weapons, although the amount and types

Conclusion

of agents that have been produced, the number and types of munitions that have been stockpiled, and the location of key research, production, and storage facilities cannot be determined with high confidence.

This assessment is heavily based on perceptions of North Korean capabilities and motivations. North Korea's large – but ageing – chemical industry is capable of producing a variety of traditional chemical weapons (CW) agents, although some imported precursors may be needed for nerve agent production. North Korea's munitions industry is capable of producing a variety of chemical weapons, such as chemical-filled artillery shells or warheads for rockets and missiles. Plausibly, Pyongyang would see the utility of chemical weapons as both a military asset for tactical battlefield use and as a strategic asset to threaten civilian casualties. Arguably, the perceived value of chemical weapons increased after the mid-1990s, when North Korea's plutonium production was frozen and its conventional forces continued to suffer from financial restrictions. North Korea denies that it has any chemical weapons, but has refused to join the Chemical Weapons Convention (CWC).

According to current South Korean official assessments, North Korea's CW programme consists of four research, eight production and seven storage sites for chemical weapons, with a stockpile of 2,500–5,000 tonnes of chemical and biological agents, including blister, nerve, choking and blood agents – as well as tear gas – which could be delivered by artillery, multiple rocket launchers, aerial bombs, FROG rockets, and *Scud* missiles. Washington currently assesses that Pyongyang can produce a range of chemical agents similar to those identified by Seoul, and estimates that North Korea possesses a 'sizeable stockpile' of these agents, which can be delivered by artillery, missiles, and aircraft, as well as Special Forces. US reports do not estimate a specific quantity of agent or munitions or discuss suspect or possible research, production, and storage sites associated with the CW programme. Prudently, US and South Korean officials assume that North Korea is prepared to use chemical weapons against military and civilian targets in a general conflict.

- There is general agreement that North Korea has conducted research and development on biological agents, but not enough information to conclude whether it has progressed to the level of agent production and weaponisation, although North Korea is most likely technically capable of both.

Compared to chemical weapons, even less is known about a possible biological weapons programme. Official US, Russian and South Korean government

reports agree that North Korea has conducted research on a variety of biological weapons agents, including anthrax, cholera, plague and smallpox, but only official South Korean sources claim that North Korea has weaponised one or two biological agents. Official US and Russian sources characterise North Korea as *capable* of producing a variety of agents, without judging that North Korea has actually produced biological weapons. Given the dearth of information, it is impossible to make a firm judgement either way. Arguably, Pyongyang might view biological weapons as relatively less significant than chemical weapons, which have more utility on the battlefield, and even less significant than nuclear weapons, which are true weapons of mass destruction.

Ballistic missiles

Assessments of North Korea's deployed short- and medium-range missiles are more certain than estimates of its efforts to develop long-range missiles capable of attacking the US with nuclear weapons.

- North Korea has produced and deployed short-range *Scud* B/C missiles (known in North Korea as the *Hwasong*-5/6), which can reach targets throughout South Korea, and medium-range *No-dong* missiles, which can reach targets throughout Japan. The exact size, disposition, and armament of these missile forces are unknown.

There is no doubt that North Korea can produce a variety of single-stage, liquid-fuelled ballistic missiles, based on *Scud* technology. However, there is little public information on the location and capabilities of missile production facilities, beyond a handful of major facilities associated with research and development, assembly, and flight-testing. With its ageing industrial base, North Korean missile production appears to be partly dependent on imports of foreign materials, equipment and components, which makes it vulnerable to supply interruptions. There is also no doubt that North Korea has deployed *Hwasong*-5/6 and *No-dong* missile units, probably organised along the lines of Soviet-style launch battalions, with four to six mobile launchers per battalion. Conservatively, we estimate a deployed force of about 120 *Hwasong*-5/6 missiles and about 40 *No-dong* missiles, but these numbers are approximate and North Korea could deploy additional missile forces if necessary. Including missiles held in reserve, official US and South Korean reports estimate that North Korea's overall ballistic missile inventory includes over 500 *Scuds* of various types and a few hundred *No-dongs*. A number of different underground bunkers, shelters, hide positions and tunnels thought to be associated with deployed missiles forces have been identified.

Presumably, Pyongyang views its short- and medium-range missile forces as a military and political asset. Armed with high-explosive or CBW warheads, missiles could serve as long-range artillery to disrupt military communications and logistics in rear areas and interdict reinforcements. As a political tool, its missile forces give North Korea a more credible threat to attack cities in South Korea and Japan with conventional or unconventional warheads, hence reinforcing deterrence and discouraging Seoul or Tokyo from pursing policies that could increase the risk of conflict. In wartime, the actual effectiveness of North Korean missiles to strike military and civilian targets would be reduced by poor accuracy, vulnerability to pre-emption, attrition of missile launchers and crews, and missile defences. However, given the number of mobile missile launchers and the variety of hide positions, some missiles would likely be launched and penetrate current defences.

- As demonstrated by the August 1998 *Taepo-dong*-1 launch, North Korea has begun to pass technological hurdles to develop multiple-stage long-range missiles, but the status of this effort cannot be accurately determined, especially since North Korea has refrained from additional flight tests since 1998.

On 31 August 1998, North Korea launched a three-stage *Taepo-dong*-1 (or *Paektusan*-1 as it is known in North Korea) rocket in an attempt to place a small satellite into orbit. Stage separation was successful (a *No-dong* first stage, *Scud* second stage and solid rocket motor third stage), but the third stage exploded and destroyed the satellite. In a ballistic missile configuration, the *Taepo-dong*-1 would provide little military utility beyond that offered by the *No-dong*, in terms of it being able to deliver a nuclear warhead to medium ranges. A more credible intercontinental range system, thought to be under development, is the *Taepo-dong*-2 (TD-2), which consists of a first stage of four clustered *No-dong* engines and a second stage of a single stage *No-dong* engine. On paper, assuming maximum capabilities, the US estimates that a two-stage TD-2 could deliver a 'nuclear weapon-sized' payload to targets in the western US. With a solid rocket motor third stage, the TD-2 is theoretically capable of delivering a 'nuclear weapon-sized' payload anywhere in the US, although accuracy would be extremely poor with known North Korean capabilities. Since 1998, the US has estimated that the TD-2 'may' be ready for testing at any time, but North Korea has refrained from additional flight tests since it agreed to a moratorium on long-range missile tests in September 1999.

While the *Taepo-dong* failed to launch a satellite into space, it succeeded in starting a debate about the status of North Korea's efforts to develop long-range missiles, which became enmeshed in longstanding disputes over National Missile Defense and the ABM Treaty. In one view, North Korea is close to developing a missile capable of attacking American cities with a nuclear warhead, which Pyongyang desires in a bid to undermine the US security relationship with South Korea and Japan. In an opposing view, North Korea's long-range missile development programme faces substantial technological obstacles and is intended more for bargaining leverage than military deployment. Both are probably true – by developing greater missile capabilities, North Korea can increase the price for abandoning its programme, while being in a stronger position to test and deploy such systems if negotiations fail. In any event, determining the actual status of North Korea's long-range missile development programme is impossible. There is some evidence that development has continued, but without flight-testing, even Pyongyang cannot be certain how close it is to achieving a successful system.

- North Korea has long been the world's leading exporter of missiles and missile technology, but sales may be declining.

Since the late 1980s, as other potential suppliers withdrew from the market, North Korea has become the world's leading missile exporter. Over the years, North Korea has sold at least several hundred *Hwasong*-5/6 or *No-dong* missiles, as well as materials, equipment, components and production technology to a range of customers, including Egypt, Iran, Libya, Pakistan, Syria, the United Arab Emirates and Yemen. In exchange, North Korea obtained cash, oil, opportunities for offshore testing and, in the case of Pakistan, nuclear technology. In recent years, however, opportunities for missile sales may have tailed off. Some of North Korea's longstanding customers, such as Iran, have nearly achieved an independent production capability, reducing their need for North Korean imports and even presenting competition to North Korean sales. Other customers, such as Pakistan, Yemen, the UAE, Egypt and, most recently, Libya, have come under political pressure from Washington to sever their missile relationship with North Korea. In addition, Pyongyang has periodically exercised caution about proceeding with missile exports that it considers too politically explosive, such as its refusal to carry out a missile deal with Iraq on the eve of the 2003 war.

The balance of conventional forces
Over the years, the conventional military balance on the Peninsula has shifted against North Korea. US and South Korean forces have modernised and strengthened their military capabilities, while North Korea's forces suffer from economic deprivation,

Conclusion

obsolete equipment, poor maintenance and inadequate training. As a result, the credibility of North Korea's threat to invade South Korea using forward-deployed forces near the Demilitarised Zone (DMZ) has diminished. With superior air power and munitions, and fighting from prepared defensive positions, US and South Korean forces stand a good chance of stopping a North Korean offensive before it could capture Seoul. In the end, North Korea could not invade the South without inviting a fatal counter attack from the US and South Korea, supported by Japan.

At the same time, Pyongyang's conventional forces are sufficiently strong to make an allied invasion to overthrow North Korea's regime an extremely unattractive option. Even with outdated equipment, poor readiness and adverse living conditions, North Korean soldiers are seen as tough fighters. With its massed artillery near the DMZ, North Korea retains the ability to inflict heavy casualties and collateral damage on allied forces and civilians – North Korean forces may not be able to seize Seoul, but they can devastate it. In theory, US forces could carry out limited pre-emptive attacks to destroy known North Korean nuclear facilities and missile emplacements, but such an attack would be unlikely to destroy all secret facilities and hidden weapons, and would risk provoking North Korean retaliatory action that could trigger a catastrophic war. The possibility that North Korea has acquired nuclear, chemical and biological weapons makes the prospect of a general war even more difficult to contemplate. With its back against the wall, the North Korean regime might take desperate and even suicidal actions.

Addressing the North Korean challenge

Because military options are unattractive, diplomacy – backed by economic and political pressures and inducements – has been the preferred instrument to restrain and dismantle North Korea's capabilities. At the same time, diplomacy requires military backing. A strong US security alliance with South Korea and Japan, and efforts to enhance allied military capabilities – including the redeployment of US forces in South Korea, continuing modernisation of South Korean forces and development of theatre missile defences – weakens North Korea's ability to employ threats and intimidation as a diplomatic instrument. The possibility of a pre-emptive strike against North Korean strategic military assets is also a diplomatic instrument. To the extent that North Korea views such an attack as a credible danger – whatever the likelihood of such a move – Pyongyang will be more inclined to avoid escalation and make compromises for a diplomatic solution. Similarly, the threat of a military blockade to enforce economic sanctions and political isolation can make a diplomatic solution more attractive as an alternative.

Over the last 25 years, a variety of diplomatic efforts have been made to address the challenges posed by North Korea's weapons programmes. These efforts have succeeded in delaying or limiting North Korea's nuclear and missile capabilities, but they have not been able to stop or eliminate them. Responding to a mixture of pressure and inducements, North Korea acceded to the NPT in December 1985 and eventually accepted IAEA inspection of its nuclear facilities and materials in April 1992. In March 1993, after refusing to cooperate with the IAEA to verify its past plutonium production, North Korea threatened to withdraw from the NPT. Other efforts were also underway during this time. In December 1991, Pyongyang and Seoul were able to conclude a broad bilateral agreement calling for a nuclear-free Korean Peninsula, only to see the agreement founder on disagreements over the frequency and intensity of inspections needed to verify rhetorical commitments. The North–South agreement remained a dead letter until it was formally renounced by Pyongyang in May 2003.

Bilateral agreements between the US and North Korea proved somewhat more effective. After North Korea threatened to withdraw from the NPT in March 1993, Washington and Pyongyang signed the Agreed Framework in October 1994. The Agreed Framework called for North Korea to freeze and eventually dismantle its plutonium production facilities and account for its plutonium stocks in exchange for interim supplies of heavy fuel oil, an alternative nuclear energy project and better relations with the US. In September 1999, the US succeeded in securing North Korean agreement to a moratorium on long-range missile tests. But the Clinton administration ran out of time to negotiate a broader agreement to freeze missile development, dismantle North Korea's existing missile force and end missile exports. The incoming Bush administration decided to pursue a 'broad agenda' seeking a comprehensive resolution of missile and other issues simultaneously rather than a stand-alone missile deal.

The Agreed Framework inhibited additional plutonium production by North Korea for nearly a decade, but it did not prevent North Korea from seeking to acquire nuclear weapons using other means. The Agreed Framework collapsed following public revelations in October 2002 that North Korea was pursuing a secret programme to produce weapons-grade uranium. In the diplomatic confrontation that followed, North Korea revived its plutonium production facilities in December 2002 and withdrew from the NPT in January 2003. It has most likely extracted some or all of the plutonium it has on hand in spent fuel – enough for a few nuclear weapons – and it has restarted the 5MW(e) reactor to produce additional plutonium. Since the collapse of the Agreed Framework, Washington has favoured multilateral diplomacy over bilateral efforts and now promotes the Six Party Talks (involving the US, Russia, China, Japan, and North and South Korea) – which have as their aim the securing of a broad-based settlement under which Pyongyang would abandon its nuclear weapons programme for security assurances and political and economic benefits.

Inevitably, the record of fitful and failed diplomacy towards North Korea over the past 25 years has taken a toll. Among all the parties involved, it has fed fatigue, suspicion and hostility. Nonetheless, the key parties remain publicly committed to dialogue and efforts to achieve a peaceful solution. The strategic challenges posed by North Korea's various weapons programmes and the implications for international and regional security are too great to be ignored or treated with passive resignation. Firstly, an unrestrained and increasingly advanced North Korean nuclear programme would deal a significant blow to the international non-proliferation regime, with North Korea setting a precedent for leaving the NPT that might be followed by others in the region and worldwide. Secondly, as North Korea's nuclear assets expand, it would make more plausible and immediate the risk that North Korean materials and capabilities could, by accident or design, find their way to unstable and hostile states or non-state actors. Thirdly, North Korea's possession of a significant nuclear force and development of long-range missile forces could undermine the basis for deterrence on the Korean Peninsula and substantially increase the military risks attending any fundamental miscalculations by Pyongyang. Finally, North Korea's weapons programmes are an increasing source of tension in East Asia – a dynamic and strategically unsettled region. As Pyongyang's nuclear efforts gather momentum and grow more visible, the likelihood will increase that countries in the region will feel the need to alter their defence and security postures to prepare for any North Korean contingency. Given the historical mutual suspicions that pervade relations among the region's major powers, essentially defensive steps are likely to detract from, rather than enhance, regional security.

Because the stakes are so high, and because other options are less attractive, diplomatic efforts are likely to continue – despite the attendant frustrations and the difficulties. The Six Party Talks face a number of very complex and contentious issues that could affect the substance of any new agreement: including the sequence and timing of coordinated steps by the different parties; the extent and type of measures necessary to verify any new agreement; and the political and economic benefits that will be provided to North Korea in exchange for disarmament. A dramatic breakthrough to resolve these issues does not appear imminent, but a continuation of the talks – as well as the intense bilateral diplomacy surrounding the formal meetings – can begin incremental progress towards a resolution of these many complex and difficult issues, a necessary move if a final agreement to address the challenges of North Korea's weapons programmes is to be reached. Finally, success in the Six Party Talks could have the additional benefit of laying the foundations for a nascent multilateral security mechanism in East Asia that would help to support peace and security throughout the region.

Notes

North Korea's Nuclear Programme

[1] Reactors use a variety of substances as moderators, which slow neutrons to the lower speeds necessary to effectively fission or split uranium. Typically, reactors fuelled by enriched uranium use light water as a moderator, while reactors fuelled by natural uranium often use heavy water or graphite as a moderator. Cooling – to help contain the heat produced by fission – can be achieved with a variety of substances, typically water or carbon dioxide.

[2] All reactors fuelled by uranium (natural or enriched) produce different isotopes of plutonium that can be used in nuclear explosions, but Plutonium 239 is the most desirable for weapons purposes. The other plutonium isotopes, such as Plutonium 238, 240 and 241 present certain complications for the design and operation of nuclear weapons because they emit heat and neutrons, which can cause premature detonation and reduce reliability. In general, the longer fuel is irradiated (that is, the higher the 'burn up'), the more these undesirable plutonium isotopes accumulate in it. 'Weapons-grade' plutonium commonly refers to plutonium with a P239 content of 90% or more, which is the grade most suitable for a first generation nuclear device.

[3] For additional technical details on the Yongbyon facilities, see Albright, D. and O'Neill, K., *Solving the North Korea Nuclear Puzzle* (Washington, DC: The Institute for Science and International Security (ISIS), 2000) and May, M. (ed) *Verifying the Agreed Framework*, (Livermore, CA: Center for Global Security Research (CGSR), Lawrence Livermore National Laboratory, April 2001).

[4] For a first hand description of the 5MW(e) reactor by an American physicist see Alvarez, Robert, 'North Korea: No bygones at Yongbyon', *Bulletin of American Scientists*, July/August 2003.

[5] The amount of radiation or 'burn up' that each fuel rod is exposed to, and, therefore, the amount of plutonium produced in each rod, depends on its location in the core. In general, rods in the centre of the core are exposed to much higher levels of irradiation than those situated on its periphery. The estimate of 7.5kg of weapons-grade plutonium produced each year is based on a discharge of the most heavily irradiated ten tonnes of fuel from the centre of the core. The remaining 40 tonnes of fuel would also contain small amounts of plutonium, but not enough to warrant reprocessing. In normal practice, fuel from the core's periphery would be moved to the centre to replace the spent fuel that has been removed and fresh fuel would be loaded into the vacant areas on the periphery.

[6] Typically, Purex reprocessing plants fail to separate 10–20% of the plutonium in the spent fuel, which remains in the waste streams. However, since the IAEA found evidence that North Korea had engaged in additional reprocessing campaigns, the claimed North Korean loss rate may not be accurate. While it is plausible that North Korea's reprocessing plant would be less efficient than Purex-based plants in the West, doubt about the actual loss rate contributes to overall uncertainty about North Korea's potential plutonium inventory.

[7] At the time, there was no hard data to test the proposition. In 1994, though, North Korea demonstrated that it could indeed unload the entire 5MW(e) reactor core in less than two months.

[8] For more technical details and alternative scenarios, see Dreicer, J., 'How Much Plutonium Could Have Been Produced in the DPRK IRT Reactor?', *Science and Global Security*, vol. 8, no. 3, 2000.

[9] Over the past decade, there have been periodic reports in the media that North Korea purchased plutonium or nuclear weapons from the former Soviet Union in the early 1990s. None of the reports can be confirmed, however, and some of the more sensational accounts provided by various North Korean defectors are generally discounted.

[10] For further details, see a series of annual reports by the IAEA Director General to the IAEA General Conference. 'Implementation of the Agreement Between the Agency and the Democratic People's Republic of Korea for the Application of Safeguards in Connection with the Treaty on the Non-Proliferation of Nuclear Weapons', GC (40)/16, 20 August 1966 and GC (41)/17, 18 August 1977. Available at www.iaea.org

[11] Sanger, D., 'North Korean Site an A-Bomb Plant, US Agencies Say', *The New York Times*, 17 August 1998.

[12] Exactly when North Korea began receiving centrifuge technology from Pakistan is unknown. Dates in public literature range from 1995–1998.

[13] Central Intelligence Agency (CIA) Report to US Congress, 19 November 2002; Warrick, J., 'US Followed the Aluminum: Pyongyang's Effort to Buy Metal was Tip to Plans', *The Washington Post*, 18 October 2002; Hibbs, M., 'Customs Intelligence Data Suggest DPRK Aimed at G-2 Type Centrifuge', *Nuclear Fuel*, vol. 28, no. 11, 26 May 2003.

[14] 'Defector Leaked Details of North Korean Nuclear Program', *Yomiuri Shimbun*, 18 December 2002.

[15] CIA report to US Congress, 19 November 2002; Frantz Douglas, 'North Korea's Nuclear Success is Doubted', *The Los Angeles Times*, 9 December 2003.

[16] Hibbs, M., 'Customs Intelligence Data Suggest DPRK Aimed at G-2 Type Centrifuge', *Nuclear Fuel*, vol. 28, no. 11, 26 May 2003.

[17] As a rule of thumb, 235 SWUs per annum are required to produce one kilogram of uranium enriched to 93% U-235. Since each G-2 machine is theoretically capable of achieving five SWUs per annum, about 50 machines of the G-2 type are needed to produce one kilogram of

weapons-grade uranium per year. Using this formula, between 1,000 and 1,250 machines would be necessary to produce 20–25kg of weapons-grade uranium a year, roughly enough for one nuclear weapon based on an implosion design. In theory, 3,500 machines could produce 75kg of weapons-grade uranium per annum, roughly enough for three weapons based on a first-generation implosion design.

[18] 'Firm Raided Over N-Part Export', *Yomiuri Shimbun*, 9 May 2003.

[19] See *Yonhap*, 19 October 2002; *Choson Ilbo*, 21 October 2002. For a comprehensive compilation of open-source information on suspect nuclear sites in North Korea, see the map on the Nuclear Threat Initiative website, www.nti.org/e_research/profiles/NK/201.html

[20] See *Korean Central News Agency*, 3 October 2003.

[21] At the time that the IAEA inspectors were expelled, the reprocessing plant's vent stack was not equipped with a filter system that could reduce emissions of krypton-85. Whether such equipment was subsequently installed after the inspectors were expelled is not known.

[22] Since the 5MW(e) reactor has historically experienced technical difficulties in operating at full power for sustained periods, a more realistic maximum production output is probably in the region of 6–7kg per year, which would yield some 4–6kg of separated plutonium, assuming a reprocessing loss of 10–30%. Presuming that each first generation implosion device requires 5–8kg of plutonium, the 5MW(e) reactor is able to produce annually about enough plutonium for one such weapon.

[23] Assuming a reprocessing loss of 10–30%, the annual yield of separated plutonium would be approximately 38.5–49.5kg per year, which could produce some five to ten nuclear weapons per year, assuming that between 5–8kg of plutonium is needed for each weapon based on a first generation implosion design.

[24] The text of the leaked KGB memo was printed in *Izvestiya* on 24 June 1994.

[25] 'Seoul Says North Korea Reprocessing Nuclear Rods', *Reuters*, 9 July 2003.

[26] See US Senate Select Committee on Intelligence, Unclassified Responses to the Questions for the Record from the Worldwide Threat Hearing of 11 February 2003, 18 August 2003, reprinted in www.fas.org/irp/congress/2003_hr/021103qfr-cia.pdf

[27] Hersh, Seymour, 'The Cold Test: The Pakistan–North Korea Nuclear Axis', *The New Yorker*, 27 January 2003.

[28] 'North Korea Holds Five Nuclear Bombs', *Agence France Presse*, 16 April 1999, quoting Japanese daily *Sankei Shimbun*. Interestingly, in November 2000, Hwang was also quoted as saying that North Korea 'has been aiming at continuing to produce nuclear weapons using uranium 235 in cooperation with a certain nation in West Asia since 1996'. See 'N. Korea

defector warns of war plans', *United Press International*, 22 April 1997.

[29] 'North Korea has Dozens of Nukes, Top Defector tells Magazine', *Agence France Presse*, 14 May 2003, quoting Japanese magazine *Gekkan Gendai*.

North Korea's Chemical and Biological Weapons Programmes

[1] Parts of this analysis are drawn from research presented at a meeting on North Korea's WMD programmes at the Carnegie Endowment for International Peace in November 2003. See Harris, Elisa D., 'North Korean Chemical and Biological Weapons Activities: Deconstructing the Threat' (Carnegie Endowment for International Peace, 7 November 2003).

[2] Oehler, Gordon, *Senate Governmental Affairs Committee Hearing to Examine Nuclear, Biological and Chemical Weapons Proliferation Threats of the 1990s* (Washington DC, 24 February 1993); Russian Foreign Intelligence Service Report, *Proliferation of Weapons of Mass Destruction* (Moscow, 1993); Republic of Korea, Ministry of National Defense, *Defence White Paper 1994* (Seoul, 1994).

[3] Bermudez, Joseph S. Jr., 'CW: North Korea's growing capabilities', *Jane's Defence Weekly*, vol. 11, no. 2, 14 January 1989, p. 54.

[4] See the Nuclear Threat Initiative, North Korea Chemical Chronology, at www.nti.org/e_research/profiles/NK/Chemical

[5] For a history of the North Korean chemical industry, see Kim Yong-yun, 'North Korean Chemical Industry', *Pukhan*, 1 December 1998, FBIS translated text, FTS19981230001322.

[6] Republic of Korea, Ministry of National Defense, *Defense White Paper 2000* (Seoul, 2000), available at www.mnd.go.kr/

[7] US Department of Defense, *Proliferation: Threat and Response*, (Washington DC, April 1996), available at www.defenselink.mil/pubs/prolif/ne_asia.html

[8] See the Nuclear Threat Initiative, North Korean Chemical Profile, www.nti.org/db/profiles/dprk/chem/over/NKC_OG0_bg.html

[9] See the Nuclear Threat Initiative, North Korea Chemical Profile, www.nti.org/db/profiles/dprk/chem/over/NKC_OG0_bg.html

[10] 'North Korea Said to have Chemical and Biological Weapons Capabilities', *United Press International*, 23 October 1992.

[11] See Kim Kyoung Soo, 'North Korea's CB Weapons: Threat and Capability', *The Korean Journal of Defense Analysis*, vol. 14, no. 1, Spring 2002, pp. 69–95.

[12] Choi Ju Hwal, North Korean Missile Proliferation Hearing before the US Senate, Governmental Affairs Subcommittee on International Security, Proliferation and Federal Services, 21 October 1997.

Notes

13 'Defector claims North completed nuclear weapons development', *KBS Radio*, 22 March 1994.

14 North Korean Missile Proliferation Hearing before the US Senate, Governmental Affairs Subcommittee on International Security, Proliferation and Federal Services, 21 October 1997.

15 'Korea Defectors', *Voice of America*, 26 February 1998.

16 See 'N. Korea defector warns of war plans', *United Press International*, 22 April 1997.

17 'North Korea: Defector says Uranium Facilities Maintained in Pakchon', *Global News Wire*, 17 October 2002, and 'A Physicist Defector's Account of North Korea's Nuke Labs', comments by Lee Wha Rang, available at www.ku.edu/~ibetext/korean-war-l/2002/10/msg00166.html

18 See the Nuclear Threat Initiative, North Korea Chemical Chronology, www.nti.org/e_research/profiles/NK/ Chemical

19 Republic of Korea, Ministry of National Defense, *Handbook on DPRK Chemical, Biological Warfare Capabilities*, 10 December 2001; and Republic of Korea, Ministry of National Defense, *Defence White Paper 2000* (Seoul, 2000), available at www.mnd.go.kr/

20 US Department of Defense, *Proliferation: Threat and Response* (Washington DC, January 2001).

21 Statement of General Thomas A. Schwartz, Commander in Chief United Nations Command/Combined Forces Command and Commander, United States Forces in Korea, before the 107th Congress, Senate Armed Services Committee, 5 March 2002, p. 8.

22 The Nuclear Threat Initiative website contains a complete catalogue of these reports, available at www.nti.org/db/profiles/dprk/chem/fac/fac_list.html. Also, See Kim Kyoung Soo, 'North Korea's CB Weapons: Threat and Capability', *The Korean Journal of Defense Analysis*, vol. 14, no. 1, Spring 2002.

23 Bermudez, Joseph S. Jr., and Richardson, Sharon A., 'The North Korean View of the Development and Production of Strategic Weapons Systems', in Sokolski, Henry (ed.), *Planning for a Peaceful Korea*, (Strategic Studies Institute, 2001), available at www.carlisle.army.mil/ssi/pubs/2001/peackora/peackora.htm

24 US Department of Defense, *Proliferation: Threat and Response*, (Washington DC, January 2001).

25 Pak Tong-sam, 'How Far has the DPRK's Development of Strategic Weapons Come?', *Pukhan*, January 1999, pp. 62-71, FBIS translated text, FTS19990121001655.

26 See the Nuclear Threat Initiative, North Korea Chemical Profile, www.nti.org/db/profiles/dprk/bio/chron/ NKB_CHGO_bg.html

27 Republic of Korea, Ministry of National Defense, *Defense White Paper 1998*, (Seoul, 1998), available at www.mnd.go.kr/

28 Russian Foreign Intelligence Service Report, *Proliferation of Weapons of Mass Destruction* (Moscow, 1993).

29 Pak Tong-sam, 'How Far Has the DPRK's Development of Strategic Weapons Come?', *Pukhan*, January 1999, pp. 62-71, FBIS translated text, FTS19990121001655.

30 US Department of Defense, *Proliferation Threat and Response*, (Washington DC, November 1997).

31 Statement by Carl W. Ford Jr., Assistant Secretary of State for Intelligence and Research before the Senate Committee on Foreign Relations Hearing on Reducing the Threat of Chemical and Biological Weapons, 19 March 2002.

32 Republic of Korea, Ministry of National Defense, *Defense White Paper 2000* (Seoul, 2000), available at www.mnd.go.kr/

33 Republic of Korea, Ministry of National Defense, *Handbook on DPRK Chemical, Biological Warfare Capabilities*, 10 December 2001.

34 US Department of Defense, *Proliferation: Threat and Response* (Washington DC, January 2001).

35 See *The New York Times*, June 1999.

36 North Korean Missile Proliferation Hearing before the United States Senate, Governmental Affairs Subcommittee on International Security, Proliferation and Federal Services, 21 October 1997.

37 Kim Kyoung Soo, 'North Korea's CB Weapons: Threat and Capability', *The Korean Journal of Defense Analysis*, vol. 14, no. 1, Spring 2002, pp. 69-95.

38 'Korea Defectors', *Voice of America*, 26 February 1998.

39 See Federation of American Scientists, www.fas.org/nuke/guide/dprk/bw/index.html

40 Bermudez, Joseph S. Jr., 'Case Study 5: North Korea', *Chemical and Biological Arms Control Institute, the Deterrence Series*, p. 12.

41 See the Nuclear Threat Initiative, North Korea Biological Chronology, www.nti.org/db/profiles/dprk/bio/ chron/NKB_CHGO_bg.html

42 US House of Representatives, Speaker's North Korea Advisory Group, *Report to the Speaker*, November 1999, available at www.fas.org/nuke/guide/dprk/nkag-report.htm

North Korea's Ballistic Missile Programme

1 See Bermudez, Joseph S. Jr., *The Armed Forces of North Korea* (New York: I.B. Taurus, 2001), pp. 240–245.

2 Lewis, John W. and Hua, Di, 'China's Ballistic Missile Programs: Technologies, Strategies, Goals', *International Security*, vol 17, fall 1992, p. 32.

3 CEP, or circular error probability, is a measure of the accuracy of the missile. It is the radius of a circle that would contain half of the impact points of a large number of missiles fired at the same point.

4 For descriptions of North Korea's missile production

facilities, see the Nuclear Threat Initiative, North Korea Country Profile, available at www.nti.org

5 See Bermudez, Joseph S. Jr., 'A History of Ballistic Missile Development in the DPRK', *Monterey Institute of International Studies, Center for Nonproliferation Studies, Occasional Paper No. 2*, November 1999, available at www.cns.miis.edu/pubs/opapers/op2/

6 'Pyongyang Pig Factory Produces Missiles', *Chosun Ilbo*, 12 February 2001.

7 Daniel A. Pinkston, interview with Kim Il Son, Monterey Institute of International Studies, Center for Nonproliferation Studies, 10 April 2001.

8 'Commercial Images Detail North Korean Missile Site', *Aviation Week & Space Technology*, 17 January, 2000.

9 Bermudez, Joseph S. Jr., 'A History of Ballistic Missile Development in the DPRK', *Monterey Institute of International Studies, Center for Nonproliferation Studies, Occasional Paper No. 2*, November, 1999, available at www.cns.miis.edu/pubs/opapers/op2/

10 'N. Korea Expanding Missile Programs', *The Washington Post*, 20 November 1998.

11 'North Korea Continues Secret Build-up Including Construction of Three Underground Bases in Rear Area', *Choson Ilbo*, 1 March 2001.

12 'North Korea Building New Missile Site, South Says', *The Washington Post*, 7 July 1999.

13 Facing a similar need for a longer-range missile during the Iran–Iraq War, Iraq modified its existing *Scud*-B missiles by extending the airframe, expanding the fuel and oxidizer tanks, modifying engine performance, and reducing payload in half, producing the *al-Hussein* missile with a range of almost 650km. The *al-Hussein* missile was aerodynamically unstable and tended to break up on re-entry, significantly reducing accuracy.

14 Wright, David and Kadyshev, Timur, 'An Analysis of the North Korean Nodong Missile', *Science and Global Security*, vol. 4, 1994, p. 129–160, available at www.princeton.edu/~globsec/publications/pdf/4_2wright.pdf

15 According to some reports, North Korea may have built a *Scud*-D variant with a 700km range and 500 kg payload.

16 McCarthy, Timothy, 'North Korean Ballistic Missile Programs: Soviet and Russian Legacies', in Michael Barletta (ed.), *WMD Threats 2001: Critical Choices for the Bush Administration*, Monterey Institute of International Studies, Center for Nonproliferation Studies, Occasional Paper No. 6, p. 9–11, available at www.cns.miis.edu/pubs/opapers/op6/index.htm

17 China's earliest ballistic missiles were heavily influenced by technology transfers from the Soviet Union. See Norris, Robert, Burrows, Andrew and Fieldhouse Richard, *British, French, and Chinese Nuclear Weapons* (Boulder: Westview Press 1994), pp. 359–362.

18 US Marine Corps, *North Korea Country Handbook* (Washington DC, 1997).

19 IISS, *The Military Balance 2003/2004* (Oxford: Oxford University Press for the IISS, 2003), p. 160.

20 US Secretary of Defense, *2000 Report to Congress: Military Situation on the Korean Peninsula*, 12 September 2000, available at www.defenselink.mil/news/Sept2000/korea09122000.html

21 Bermudez, Joseph S. Jr., *The Armed Forces of North Korea* (New York: I.B. Taurus, 2001).

22 For details or reported missile deployment sites, see the Nuclear Threat Initiative, North Korea Country Profile, www.nti.org

23 'New Missile Reported in North Korea; California is within its Range, a Defector tells South Koreans', *The Herald Tribune*, 19 February 2000, and 'A Tale of Two Defectors', *The Washington Times*, 16 March 2000.

24 'Defector: North Korea has capability to put satellite in orbit', *Associated Press*, 7 September 1998.

25 'South Korea: Defector Provides 'Unique' Opportunity for Arms Information', *Yonhap*, 27 August 1997.

26 'North Korea aims missiles at Tokyo: defector', *United Press International*, 6 June 1997, and 'North Korea arms target Tokyo: defector', *Courier Mail*, 10 June 1997.

27 Hearing on Drugs, Counterfeiting, and Weapons Proliferation: The North Korean Connection, the Financial Management, the Budget, and International Security, Subcommittee of the US Senate Governmental Affairs Committee, 20 May 2003.

28 'Report: North Korea would not launch pre-emptive attack on Japan, defector says', *Associated Press*, 28 May 2003; 'North Korea Imports 90 Percent of Missile Parts from Japan says Defector', BBC, 16 May 2003.

29 'North Korea – Defector Claims Pyongyang has Dozens of Nukes', *IAC (SM) Newsletter Database*, 15 May 2003, and 'North Korea has Dozens of Nukes, Top Defector Tells Magazine', *Agence France Presse*, 14 May 2003.

30 'North Korea has Pak-Made Nukes, Says Defector', *The Economic Times of India*, 28 November 2002.

31 'Defector says 1991 missile plant explosion killed 200', *BBC Summary of World Broadcasts*, 21 March 1994.

32 North Korean Missile Proliferation Hearing before the US Senate, Governmental Affairs Subcommittee on International Security, Proliferation, and Federal Services, 21 October 1997.

33 Ko Young Hwan, North Korean Missile Proliferation Hearing before the US Senate, Governmental Affairs Subcommittee on International Security, Proliferation, and Federal Services, 21 October 1997.

34 See the Nuclear Threat Initiative, www.nti.org/e_research/profiles/NK/Missile/64_750.html

35 'Ballistic Missile Program, Defense and Foreign Affairs', *Strategic Policy*, December, 1999.

Notes

36 www.nti.org/e_research/profiles/NK/Missile/ 64_750.html

37 www.nti.org/e_research/profiles/NK/Missile/ 64_754.html

38 North Korean defector's press conference, *KBS Television*, 11 November 1996.

39 'North Korea's Scuds said capable of carrying chemicals', *Kyodo News International*, 28 April 1994.

40 There may be some variations on the range and payload of the different *No-dong* models. Some accounts report a *No-dong*-1 and *No-dong*-2.

41 See IISS, *Iraq's Weapons of Mass Destruction: A Net Assessment* (London, 2002), pp 59–60.

42 See Monterey Institute of International Studies, Center for Nonproliferation Studies, Chronology of North Korea's Missile Trade and Developments: 1992–93, available at cns.miis.edu/iiop

43 Pakistan *Ghauri* missiles were successfully tested in April 1998, April 1999, and May 2002. Iran's *Shahab*-3 tests in July 1998 and September 2000 are reported to have failed, while tests in July 2000 and May 2002 are believed to have been successful.

44 Wright, David and Kadyshev, Timur, 'An Analysis of the North Korean Nodong Missile', *Science and Global Security*, vol. 4, 1994, available at www.princeton.edu/ ~globsec/publications/ pdf/4_2wright.pdf

45 Republic of Korea, Ministry of National Defense, *Defence White Paper, 2000* (Seoul, 2000), available at www.mnd.go.kr/

46 See IISS, *The Military Balance 2003/2004* (Oxford: Oxford University Press for the IISS, 2003), p. 160; 'N. Korea has Up to 750 Ballistic Missiles: U.S. Source', *Jiji Press Ticker Service*, 12 May 2003.

47 For further information on reported missile deployment sites, see the Nuclear Threat Initiative, North Korean Country Profile, www.nti.org

48 Smith, Jeffrey R., 'N. Korea and the Bomb: High Tech Hide-and-Seek; US Intelligence Key In Detecting Deception', *The Washington Post*, 18 March 1994.

49 *Paektusan* means Mount Paektu (literally White Head Mountain), the highest mountain in Korea and redoubt of Kim Il-Sung during his guerrilla struggle against the Japanese in the late 1930s and legendary birth place of Kim Jong Il.

50 Details of the launch, including the timing of the staging of the booster, were given in the initial North Korean press reports of the launch, 'Successful Launch of First Satellite in DPRK', *Korean Central News Agency*, 4 September 1998, available at www.kcna.co.jp/item/1998/9809/news09/04.htm#1

51 National Intelligence Council, *Foreign Missile Developments and the Ballistic Missile Threat Through 2015*, September 1999, available at www.cia.gov/nic/pubs/other_products/foreign_ missile_developments.htm

52 National Intelligence Estimate, *Emerging Missile Threats to North America During the Next 15 Years*, November 1995, available at www.fas.org/spp/ starwars/offdocs/nie9519.htm

53 Report of the Commission to Assess the Ballistic Missile Threat to the United States, 15 July 1998. Available at www.access.gpo.gov/su_docs/ newnote.html

54 National Intelligence Estimate, *Emerging Missile Threats to North America During the Next 15 Years*, November 1995, available at www.fas.org/spp/ starwars/offdocs/nie9519.htm

55 The NIEs do not specify what 'several hundred kilograms' means, but presumably it is meant to include at least 500kg, the standard defined by the Missile Technology Control Regime (MTCR) as the minimum necessary for a first generation nuclear weapon and heat shield. Of course, it is not known whether North Korea is capable of producing a nuclear warhead of this class, and a larger payload would significantly reduce the range of the missile. For example, a missile of the size of a TD-2 that could carry a 500kg payload 12,000km could reach only 7,000–8,000km with a 1,000kg payload.

56 National Intelligence Council, *Foreign Missile Developments*, December 2001.

57 See, for example, Sessler, Andrew, *Countermeasures : A Technical Evaluation of the Operational Effectiveness of the Planned US National Defense System* (Cambridge, MA: Union of Concerned Scientists and MIT Security Studies Program, 2000).

58 Federation of American Scientists, No-dong Launch Facility, available at www.fas.org/nuke/guide/ dprk/facility/Nodong.htm; Bermudez, Joseph Jr., 'North Korea's Musudan-Ri Launch Facility', available at www.cdiss.org/spec99aug.htm

59 See Monterey Institute of International Studies, Center for Nonproliferation Studies, Chronology of North Korea's Missile Trade and Developments: 1999–2002, available at cns.miis.edu.htm

60 See 'Russia: N. Korea Unable to Advance in Missile Development', *Middle East Newsline*, vol. 3, no. 248, 26 June 2001, www.menesline.com

61 See Kniazkov, Maxim, 'North Korea has new intermediate range missile', *Agence France Presse*, 11 September 2003; 'Missile Watch as N. Korea turns 55', CNN, 8 September 2003.

62 Cochran,Thomas, Arkin, William, Norris, Robert and Sands, Jeffrey, *Soviet Nuclear Weapons* (New York: Ballinger, 1989), pp. 143–144.

63 The usual estimate of the number of missiles exported by North Korea is 400, but this is only a rough estimate that dates back to 1996 and so it is probably out of date. For details on North Korea's missile exports to particular countries, see the Nuclear Threat Initiative, North Korea Country Profile, www.nti.org

64 'Pakistan's Missile 'Was a Nodong'', *Jane's Missiles & Rockets*, vol. 2, no. 5, May 1998, pp. 1–2.

65 Yi Kyo-kwan, 'How Does North Korea Export Missiles?', *Chosun Ilbo*, 5 March 2002, in *DPRK Said to Export Body, Main Parts of Missiles Separately*, FBIS translated text, KPP20020305000112; Lee Kyo-kwan, 'NK Missile Exports Diversified in Technique', *Chosun Ilbo*, 7 March 2002, available at nk.chosun.com/english

Conventional Military Balance

1 Hodge, Homer T., 'North Korea's Military Strategy', *Parameters*, *US Army War College Quarterly*, Spring 2003, available at carlisle-www.army.mil/usawc/Parameters/03spring/hodge.pdf; Kongdan, Oh and Hassig, Ralph C., *North Korea Through the Looking Glass* (Washington DC: Brookings, 2000); Oberdorfer, Don, *The Two Koreas* (Reading: Mass: Addison-Wesley, 1997).

2 Quantities for North Korean military formations and its inventory of weapons are not known exactly. This chapter draws from the IISS, *The Military Balance 2003/2004* (Oxford: Oxford University Press for the IISS, 2003), for slightly different figures, see Republic of Korea, Ministry of National Defense, *Defense White Paper 2000* (Seoul, 2000), available at www.mnd.go.kr/; and US Department of Defense, *2000 Report to Congress: Military Situation on the Korean Peninsula* (Washington DC, 12 September 2000), available at www.defenselink.mil/news/Sep2000/korea09122000.html

3 See O'Hanlon, Michael, 'Stopping a North Korean Invasion: Why Defending South Korea is Easier than the Pentagon Thinks', *International Security*, vol. 22, no. 4, Spring 1998.

4 Office of Naval Intelligence, *Worldwide Submarine Proliferation in the Coming Decade* (Washington DC: US Department of Defense, May 1995).

5 Quantities for ROK military structure and inventory are drawn from the IISS, *The Military Balance 2003/2004* (Oxford: Oxford University Press for the IISS, 2003).

6 US Department of Defense, *2000 Report to Congress: Military Situation on the Korean Peninsula* (Washington DC, 12 September 2000), p. 8, available at www.defenselink.mil/news/Sep2000/korea09122000.html

7 Statement of General Thomas A. Schwartz, Commander in Chief, United Nations Command/Combined Forces Command, and Commander, United States Forces Korea, before the Senate Armed Services Committee, 5 March 2002, available at www.defenselink.mil

8 See O'Hanlon, Michael, 'Stopping a North Korean Invasion: Why Defending South Korea is Easier than the Pentagon Thinks', *International Security*, vol. 22, no. 4, Spring 1998.

9 See, for example, Mangold, Tom and Goldberg, Jeff, *Plague Wars: The Terrifying Reality of Biological Warfare* (London: Macmillan, 1999), pp. 325–327, for an account of a US military exercise in 1996 that began with the projected use of biological weapons, delivered by North Korean Special Forces.

10 Statement of General Thomas A. Schwartz, Commander in Chief, United Nations Command/Combined Forces Command, and Commander, United States Forces Korea, before the Senate Armed Services Committee, 5 March 2002, available at www.defenselink.mil

TREATY ON THE NON-PROLIFERATION OF NUCLEAR WEAPONS

The States concluding this Treaty, hereinafter referred to as the "Parties to the Treaty",

Considering the devastation that would be visited upon all mankind by a nuclear war and the consequent need to make every effort to avert the danger of such a war and to take measures to safeguard the security of peoples,

Believing that the proliferation of nuclear weapons would seriously enhance the danger of nuclear war,

In conformity with resolutions of the United Nations General Assembly calling for the conclusion of an agreement on the prevention of wider dissemination of nuclear weapons,

Undertaking to co-operate in facilitating the application of International Atomic Energy Agency safeguards on peaceful nuclear activities,

Expressing their support for research, development and other efforts to further the application, within the framework of the International Atomic Energy Agency safeguards system, of the principle of safeguarding effectively the flow of source and special fissionable materials by use of instruments and other techniques at certain strategic points,

Affirming the principle that the benefits of peaceful applications of nuclear technology, including any technological by-products which may be derived by nuclear-weapon States from the development of nuclear explosive devices, should be available for peaceful purposes to all Parties to the Treaty, whether nuclear-weapon or non-nuclear-weapon States,

Convinced that, in furtherance of this principle, all Parties to the Treaty are entitled to participate in the fullest possible exchange of scientific information for, and to contribute alone or in co-operation with other States to, the further development of the applications of atomic energy for peaceful purposes,

Declaring their intention to achieve at the earliest possible date the cessation of the nuclear arms race and to undertake effective measures in the direction of nuclear disarmament,

Urging the co-operation of all States in the attainment of this objective,

Recalling the determination expressed by the Parties to the 1963 Treaty banning nuclear weapon tests in the atmosphere, in outer space and under water in its Preamble to seek to achieve the discontinuance of all test explosions of nuclear weapons for all time and to continue negotiations to this end,

Desiring to further the easing of international tension and the strengthening of trust between States in order to facilitate the cessation of the manufacture of nuclear weapons, the liquidation of all their existing stockpiles, and the elimination from national arsenals of nuclear weapons and the means of their delivery pursuant to a Treaty on general and complete disarmament under strict and effective international control,

Recalling that, in accordance with the Charter of the United Nations, States must refrain in their international relations from the threat or use of force against the territorial integrity or political independence of any State, or in any other manner inconsistent with the Purposes of the United Nations, and that the establishment and maintenance of international peace and security are to be promoted with the least diversion for armaments of the world's human and economic resources,

Have agreed as follows:

Article I
Each nuclear-weapon State Party to the Treaty undertakes not to transfer to any recipient whatsoever nuclear weapons or other nuclear explosive devices or control over such weapons or explosive devices directly, or indirectly; and not in any way to assist, encourage, or induce any non-nuclear-weapon State to manufacture or otherwise acquire nuclear weapons or other nuclear explosive devices, or control over such weapons or explosive devices.

Article II
Each non-nuclear-weapon State Party to the Treaty undertakes not to receive the transfer from any transferor whatsoever of nuclear weapons or other nuclear explosive devices or of control over such weapons or explosive devices directly, or indirectly; not to manufacture or otherwise acquire nuclear weapons or other nuclear explosive devices; and not to seek or receive any assistance in the manufacture of nuclear weapons or other nuclear explosive devices.

Article III
1. Each non-nuclear-weapon State Party to the Treaty undertakes to accept safeguards, as set forth in an agreement to be negotiated and concluded with the International Atomic Energy Agency in accordance with the Statute of the International Atomic Energy Agency and the Agency's safeguards system, for the exclusive purpose of verification of the fulfilment of its obligations assumed under this Treaty with a view to preventing diversion of nuclear energy from peaceful uses to nuclear weapons or other nuclear explosive devices. Procedures for the safeguards required by this Article shall be followed with respect to source or special fissionable material whether it is being produced, processed or used in any principal nuclear facility or is outside any such facility. The safeguards required by this Article shall be applied on all source or special fissionable material in all peaceful nuclear activities within the territory of such State, under its jurisdiction, or carried out under its control anywhere.

2. Each State Party to the Treaty undertakes not to provide: (a) source or special fissionable material, or (b) equipment or material especially designed or prepared for the processing, use or production of special fissionable material, to any non-nuclear-weapon State for peaceful purposes, unless the source or special fissionable material shall be subject to the safeguards required by this Article.

3. The safeguards required by this Article shall be implemented in a manner designed to comply with Article IV of this Treaty, and to avoid hampering the economic or technological development of the Parties or international co-operation in the field of peaceful nuclear activities, including the international exchange of nuclear material and equipment for the processing, use or production of nuclear material for peaceful purposes in accordance with the provisions of this Article and the principle of safeguarding set forth in the Preamble of the Treaty.

4. Non-nuclear-weapon States Party to the Treaty shall conclude agreements with the International Atomic Energy Agency to meet the requirements of this Article either individually or together with other States in accordance with the Statute of the International Atomic Energy Agency. Negotiation of such agreements shall commence within 180 days from the original entry into force of this Treaty. For States depositing their instruments of ratification or accession after the 180-day period, negotiation of such agreements shall commence not later than the date of such deposit. Such agreements shall enter into force not later than eighteen months after the date of initiation of negotiations.

Article IV
1. Nothing in this Treaty shall be interpreted as affecting the inalienable right of all the Parties to the Treaty to develop research, production and use of nuclear energy for peaceful purposes without discrimination and in conformity with Articles I and II of this Treaty.

2. All the Parties to the Treaty undertake to facilitate, and have the right to participate in, the fullest possible exchange of equipment, materials and scientific and technological information for the peaceful uses of nuclear energy. Parties to the Treaty in a position to do so shall also co-operate in contributing alone or together with other States or international organizations to the further development of the applications of nuclear energy for peaceful purposes, especially in the territories of non-nuclear-weapon States Party to the Treaty, with due consideration for the needs of the developing areas of the world.

Article V
Each Party to the Treaty undertakes to take appropriate measures to ensure that, in accordance with this Treaty, under appropriate international observation and through appropriate international procedures, potential benefits from any peaceful applications of nuclear explosions will be made available to non-nuclear-weapon States Party to the Treaty on a nondiscriminatory basis and that the charge to such Parties for the explosive devices used will be as low as possible and exclude any charge for research and development. Non-nuclear-weapon States Party to

Treaty on the non-proliferation of nuclear weapons

the Treaty shall be able to obtain such benefits, pursuant to a special international agreement or agreements, through an appropriate international body with adequate representation of non-nuclear-weapon States. Negotiations on this subject shall commence as soon as possible after the Treaty enters into force. Non-nuclear-weapon States Party to the Treaty so desiring may also obtain such benefits pursuant to bilateral agreements.

Article VI
Each of the Parties to the Treaty undertakes to pursue negotiations in good faith on effective measures relating to cessation of the nuclear arms race at an early date and to nuclear disarmament, and on a treaty on general and complete disarmament under strict and effective international control.

Article VII
Nothing in this Treaty affects the right of any group of States to conclude regional treaties in order to assure the total absence of nuclear weapons in their respective territories.

Article VIII

1. Any Party to the Treaty may propose amendments to this Treaty. The text of any proposed amendment shall be submitted to the Depositary Governments which shall circulate it to all Parties to the Treaty. Thereupon, if requested to do so by one-third or more of the Parties to the Treaty, the Depositary Governments shall convene a conference, to which they shall invite all the Parties to the Treaty, to consider such an amendment.

2. Any amendment to this Treaty must be approved by a majority of the votes of all the Parties to the Treaty, including the votes of all nuclear-weapon States Party to the Treaty and all other Parties which, on the date the amendment is circulated, are members of the Board of Governors of the International Atomic Energy Agency. The amendment shall enter into force for each Party that deposits its instrument of ratification of the amendment upon the deposit of such instruments of ratification by a majority of all the Parties, including the instruments of ratification of all nuclear-weapon States Party to the Treaty and all other Parties which, on the date the amendment is circulated, are members of the Board of Governors of the International Atomic Energy Agency. Thereafter, it shall enter into force for any other Party upon the deposit of its instrument of ratification of the amendment.

3. Five years after the entry into force of this Treaty, a conference of Parties to the Treaty shall be held in Geneva, Switzerland, in order to review the operation of this Treaty with a view to assuring that the purposes of the Preamble and the provisions of the Treaty are being realised. At intervals of five years thereafter, a majority of the Parties to the Treaty may obtain, by submitting a proposal to this effect to the Depositary Governments, the convening of further conferences with the same objective of reviewing the operation of the Treaty.

Article IX

1. This Treaty shall be open to all States for signature. Any State which does not sign the Treaty before its entry into force in accordance with paragraph 3 of this Article may accede to it at any time.

2. This Treaty shall be subject to ratification by signatory States. Instruments of ratification and instruments of accession shall be deposited with the Governments of the United Kingdom of Great Britain and Northern Ireland, the Union of Soviet Socialist Republics and the United States of America, which are hereby designated the Depositary Governments.

3. This Treaty shall enter into force after its ratification by the States, the Governments of which are designated Depositaries of the Treaty, and forty other States signatory to this Treaty and the deposit of their instruments of ratification. For the purposes of this Treaty, a nuclearweapon State is one which has manufactured and exploded a nuclear weapon or other nuclear explosive device prior to 1 January, 1967.

4. For States whose instruments of ratification or accession are deposited subsequent to the entry into force of this Treaty, it shall enter into force on the date of the deposit of their instruments of ratification or accession.

5. The Depositary Governments shall promptly inform all signatory and acceding States of the date of each signature, the date of deposit of each instrument of ratification or of accession, the date of the entry into force of this Treaty, and the date of receipt of any requests for convening a conference or other notices.

Treaty on the non-proliferation of nuclear weapons

6. This Treaty shall be registered by the Depositary Governments pursuant to Article 102 of the Charter of the United Nations.

Article X
1. Each Party shall in exercising its national sovereignty have the right to withdraw from the Treaty if it decides that extraordinary events, related to the subject matter of this Treaty, have jeopardized the supreme interests of its country. It shall give notice of such withdrawal to all other Parties to the Treaty and to the United Nations Security Council three months in advance. Such notice shall include a statement of the extraordinary events it regards as having jeopardized its supreme interests.

2. Twenty-five years after the entry into force of the Treaty, a conference shall be convened to decide whether the Treaty shall continue in force indefinitely, or shall be extended for an additional fixed period or periods. This decision shall be taken by a majority of the Parties to the Treaty.

Article XI
This Treaty, the English, Russian, French, Spanish and Chinese texts of which are equally authentic, shall be deposited in the archives of the Depositary Governments. Duly certified copies of this Treaty shall be transmitted by the Depositary Governments to the Governments of the signatory and acceding States.

IN WITNESS WHEREOF the undersigned, duly authorised, have signed this Treaty.

DONE in triplicate, at the cities of London, Moscow and Washington, the first day of July, one thousand nine hundred and sixty-eight.

Entered into force: 5 March 1970. Source: U.N.T.S. No. 10485, vol. 729, pp. 169–175.

JOINT DECLARATION ON THE DENUCLEARIZATION OF THE KOREAN PENINSULA

Entry into force on February 19, 1992

South and North Korea,

In order to eliminate the danger of nuclear war through the denuclearization of the Korean peninsula, to create conditions and an environment favourable to peace and the peaceful unification of Korea, and thus to contribute to the peace and security of Asia and the world,

Declare as follows;

1. South and North Korea shall not test, manufacture, produce, receive, possess, store, deploy or use nuclear weapons.

2. South and North Korea shall use nuclear energy solely for peaceful purposes.

3. South and North Korea shall not possess nuclear reprocessing and uranium enrichment facilities.

4. In order to verify the denuclearization of the Korean peninsula, South and North Korea shall conduct inspections of particular subjects chosen by the other side and agreed upon between the two sides, in accordance with the procedures and methods to be determined by the South-North Joint Nuclear Control Commission.

5. In order to implement this joint declaration, South and North Korea shall establish and operate a South-North Joint Nuclear Control Commission within one month of the entry into force of this joint declaration;

6. This joint declaration shall enter into force from the date the South and the North exchange the appropriate instruments following the completion of their respective procedures for bringing it into effect.

Chung Won-shik
Chief Delegate of the South delegation
to the South-North High-Level Negotiations
Prime Minister of the Republic of Korea

Yon Hyong-muk
Head of the North delegation to the
South-North High-Level Negotiations
Premier of the Administration Council
of the Democratic People's Republic of Korea

AGREED FRAMEWORK BETWEEN THE UNITED STATES OF AMERICA AND THE DEMOCRATIC PEOPLE'S REPUBLIC OF KOREA, GENEVA, OCTOBER 21, 1994

Delegations of the Governments of the United States of America (US) and the Democratic People's Republic of Korea (DPRK) held talks in Geneva from September 23 to October 21, 1994, to negotiate an overall resolution of the nuclear issue on the Korean Peninsula.

Both sides reaffirmed the importance of attaining the objectives contained in the August 12, 1994 agreed statement between the US and the DPRK and upholding the principles of the June 11, 1993 joint statement of the US and the DPRK to achieve peace and security on a nuclear-free Korean Peninsula. The US and DPRK decided to take the following actions for the resolution of the nuclear issue:

I. Both sides will cooperate to replace the DPRK's graphite-moderated reactors and related facilities with light-water reactor (LWR) power plants.

1) In accordance with the October 20, 1994 letter of assurance from the US President, the US will undertake to make arrangements for the provision to the DPRK of a LWR project with a total generating capacity of approximately 2,000MW(E) by a target date of 2003.
   - The US will organize under its leadership an international consortium to finance and supply the LWR project to be provided to the DPRK. The US, representing the international consortium, will serve as the principal point of contact with the DPRK for the LWR project.
   - The US, representing the consortium, will make best efforts to secure the conclusion of a supply contract with the DPRK within six months of the date of this document for the provision of the LWR project. Contract talks will begin as soon as possible after the date of this document.
   - As necessary, the US and the DPRK will conclude a bilateral agreement for cooperation in the field of peaceful uses of nuclear energy.

2) In accordance with the October 20, 1994 letter of assurance from the US President, the US, representing the consortium, will make arrangements to offset the energy foregone due to the freeze of the DPRK's graphite-moderated reactors and related facilities, pending completion of the first LWR unit.
   - Alternative energy will be provided in the form of heavy oil for heating and electricity production.
   - Deliveries of heavy oil will begin within three months of the date of this document and will reach a rate of 500,000 tons annually, in accordance with an agreed schedule of deliveries.

3) Upon receipt of US assurances for the provision of LWR's and for arrangements for interim energy alternatives, the DPRK will freeze its graphite-moderated reactors and related facilities and will eventually dismantle these reactors and related facilities.
   - The freeze on the DPRK's graphite-moderated reactors and related facilities will be fully implemented within one month of the date of this document. During this one-month period, and throughout the freeze, the International Atomic Energy Agency (IAEA) will be allowed to monitor this freeze, and the DPRK will provide full cooperation to the IAEA for this purpose.
   - Dismantlement of the DPRK's graphite-moderated reactors and related facilities will be completed when the LWR project is completed. – The US and DPRK will cooperate in finding a method to store safely the spent fuel from the 5MW(E) experimental reactor during the construction of the LWR project, and to dispose of the fuel in a safe manner that does not involve reprocessing in the DPRK.

4) As soon as possible after the date of this document, US and DPRK experts will hold two sets of experts talks.
   - At one set of talks, experts will discuss issues related to alternative energy and the replacement of the graphite-moderated reactor program with the LWR project.
   - At the other set of talks, experts will discuss specific arrangements for spent fuel storage and ultimate disposition.

Agreed framework between the United States of America and the Democratic People's Republic of Korea, Geneva, October 21, 1994

II. The two sides will move toward full normalization of political and economic relations.

1) Within three months of the date of this document, both sides will reduce barriers to trade and investment, including restrictions on telecommunications services and financial transactions.
2) Each side will open a liaison office in the other's capital following resolution of consular and other technical issues through expert level discussions.
3) As progress is made on issues of concern to each side, the US and DPRK will upgrade bilateral relations to the ambassadorial level.

III. Both sides will work together for peace and security on a nuclear-free Korean Peninsula.

1) The US will provide formal assurances to the DPRK, against the threat or use of nuclear weapons by the US
2) The DPRK will consistently take steps to implement the North-South Joint Declaration on the Denuclearization of the Korean Peninsula.
3) The DPRK will engage in North-South dialogue, as this agreed framework will help create an atmosphere that promotes such dialogue.

IV. Both sides will work together to strengthen the international nuclear non-proliferation regime.

1) The DPRK will remain a party to the Treaty on the Non-Proliferation of Nuclear Weapons (NPT) and will allow implementation of its Safeguards Agreement under the Treaty.
2) Upon conclusion of the supply contract for the provision of the LWR project, ad hoc and routine inspections will resume under the DPRK's Safeguards Agreement with the IAEA with respect to the facilities not subject to the freeze. Pending conclusion of the supply contract, inspections required by the IAEA for the continuity of safeguards will continue at the facilities not subject to the freeze.
3) When a significant portion of the LWR project is completed, but before delivery of key nuclear components, the DPRK will come into full compliance with its Safeguards Agreement with the IAEA (INFCIRC/403), including taking all steps that may be deemed necessary by the IAEA, following consultations with the Agency with regard to verifying the accuracy and completeness of the DPRK's initial report on all nuclear material in the DPRK.

Kang Sok Ju
Head of the Delegation of the Democratic
People's Republic of Korea, First Vice-Minister
of Foreign Affairs of the Democratic People's Republic of Korea

Robert L. Gallucci
Head of the Delegation of the United States of America, Ambassador at Large of the United States of America

Source: www.kedo.org/pdfs/AgreedFramework.pdf